SIR ISAAC NEWTON'S

ENUMERATION OF LINES OF THE THIRD ORDER.

LONDON:
Printed by G. BARCLAY, Castle St. Leicester Sq.

SIR ISAAC NEWTON'S

ENUMERATION OF LINES OF THE THIRD ORDER,

GENERATION OF CURVES BY SHADOWS,

ORGANIC DESCRIPTION OF CURVES,

AND

CONSTRUCTION OF EQUATIONS BY CURVES.

TRANSLATED FROM THE LATIN.

WITH NOTES AND EXAMPLES.

BY

C. R. M. TALBOT, M.P. F.R.S.

LONDON:

H. G. BOHN, YORK STREET, COVENT GARDEN.

MDCCCLX.

PREFACE.

It is the object of the present volume to place within reach of the student a work to which he is frequently referred in the course of his reading, but which he will seldom have the opportunity of consulting, unless he happens to be possessed of the quarto edition of Sir I. Newton's work; for the "Enumeratio linearum tertii ordinis," originally printed as an appendix to the treatise on Optics, and subsequently in the "Opuscula," has not hitherto been published in a separate form.

In 1717, Mr. Stirling, of Balliol College, Oxford, published his commentary on this work, entitled "Illustratio tractatus D. Newtoni de enumeratione linearum tertii ordinis;" but this book, though reprinted at Paris in 1797, is now scarce, and beyond the reach of the general reader.

To have published a translation of the mere text of Newton's treatise, without the addition of any explanation or illustration, would have been a barren and useless undertaking, for its style is concise and abrupt almost to obscurity, and its enunciations are unaccom-

panied by any proofs. The Abbé de Gua, in his
"Usage de l'Analyse de Descartes," in remarking on
the difficulty of following the thread of the argument
made use of by Newton, writes thus :—" Ce géomètre
dont tous les ouvrages portent un caractère singulier de
sublimité, paroit en particulier dans celui-ci s'être élevé
à une hauteur immense, à laquelle toute autre génie
moins pénétrant et moins fort que le sien, auroit tenté
vainement d'atteindre : mas la route qu'il a tenue dans
une entreprise si difficile, se dérobe aux yeux de ceux
qui apperçoivent avec étonnement le degré d'élévation
auquel il est parvenu. On doit en excepter quelques
legères traces qu'il a eu soin de laisser sur son
passage, aux endroits qui avoient mérité qu'il s'y ar-
rétât plus long tems. Ces endroits, au reste, sont
presque toujours assez distants les uns des autres. Si
l'on se propose donc de suivre la même carrière, on
est obligé se guider soi-même dans de longs inter-
valles." And Cramer has gone so far as to accuse
Newton of purposely withholding the demonstrations
of the method employed by him in this work ; for, in
speaking of it, he says, " Il est facheux que M.
Newton se soit contenté d'étaler ses découvertes sans y
joindre les démonstrations, et qu'il ait préféré le plaisir
de se faire admirer à celui d'instruire."

 There is, however, little doubt that Newton's object
in hastily publishing this treatise, was to vindicate the

priority of his own discoveries from the piracy of writers who had appropriated some of his theorems; and, in the advertisement prefixed to the " Optics," he complains that " having lent out his manuscript containing such theorems, he had since met with some things copied out of it, and therefore took occasion to publish the tract." Nor can it be doubted that the work of Stirling, which supplies the omitted demonstrations, was sanctioned and assisted by Newton himself, with whom the author was on terms of intimate friendship.

But although the criticism of the French mathematicians would appear to be ill founded, it cannot be denied that to those who have no previous familiarity with the subjects treated in the " Enumeratio," some explanation and illustration is wanted in order to enable them to arrive at the author's meaning; and it is for this reason, and with a view of supplying the desideratum, that notes to elucidate the text have been added in the present volume.

Occasion has also been taken in the following pages to give a short account of the application of Newton's analytical parallelogram to the investigation of curve-lines,—a method which seems to have undeservedly fallen into disuse. The writer fully concurs in the truth of the remark of Stirling, " Hâc methodo universali procedendo, scilicet argumentando à naturis

æquationum, patescunt non solùm sectionum coni proprietates, quas tanto labore adinvenerunt Veteres, et
tot ambagibus demonstratas dederunt, idque methodo
quæ ad alias curvas extendi nequit; sed et proprietates
curvarum omnium ordinum superiorum."

A few examples and problems relating to lines of
the third order have been added at the end of the
volume, in order to render the work more useful to
the student.

CONTENTS.

AN ENUMERATION OF LINES OF THE THIRD ORDER.

Section I.

The Orders of Lines.

Geometrical lines are best divided into orders, according to the dimensions of the equation expressing the relation between absciss and ordinate, or, which is the same thing, according to the number of points in which they can be cut by a straight line. So that a line of the first order will be a straight line; those of the second or quadratic order will be conic sections and the circle; and those of the third or cubic order will be the cubic Parabola, the Neilian Parabola, the Cissoïd of the ancients, and others we are about to describe. A curve of the first genus (since straight lines are not to be reckoned among curves) is the same as a line of the second order, and a curve of the second genus is the same as a line of the third order. And a line of the infinitesimal order is one which a straight line may cut in an infinite number of points, such as the spiral, cycloïd, quadratrix, and every line generated by the infinitely continued rotations of a radius.

SECTION II.

The Properties of Conic Sections are analogous to those of Curves of higher orders.

The chief properties of conic sections have been much treated of by geometers, and the properties of curves of the second and higher genera are very similar to them, as will be shown in the following enumeration of their principal properties:—

1. *Of Curves of the Second Genus, their Ordinates, Diameters, Vertices, Centres, and Axes.*

If parallel straight lines terminated by the curve be drawn in a conic section, the straight line bisecting two of them will bisect all the others, and is called the diameter, and the bisected lines are called ordinates to the diameter, and the intersection of all the diameters is the centre; the intersection of the diameter with the curve is called the vertex; that diameter, whose ordinates are rectangular to it, being called the axis. In like manner, in curves of the second genus, if any two parallel straight lines are drawn, meeting the curve in three points; the straight line which cuts these parallel lines, so that the sum of the two segments meeting the curve on one side of the secant, equals the third segment meeting the curve on the other side of the secant, will cut in the same ratio all lines parallel to these, provided they also meet the curve in three points; that is, so that the sum of the two parts on one side of the secant, shall equal the third part on the other side. These three parts, thus equal, may be called ordinates, and the secant or cutting line to which the ordinates are applied, the diameter; the intersection of diameter and curve, the vertex; and the intersection of two diameters, the centre. The diameter having rectangular ordinates, if any exist, may also be called an axis; and where all the diameters meet in a point, that point will be the general centre.

2. *Of Asymptotes and their Properties.*

A hyperbola of the first genus will have two asymptotes; that of the second genus will have three; of the third, four, and no more; and so on for the rest. And as the segments of any straight line intercepted between the conic hyperbola and its asymptotes on each side are equal, so in hyperbolas of the second genus, if any straight line be drawn cutting both the curve and its three asymptotes in three points, the sum of those two segments of the secant, which are drawn from any two asymptotes on the same side to two points of the curve, will be equal to the third part, which is drawn from the third asymptote on the contrary side, to the third point in the curve.

3. *Of Latera Recta and Transversa.*

And in like manner, as in the non-parabolic conic sections, the square of the ordinate—*i. e.* the rectangle contained by the ordinates on each side of the diameter—is to the rectangle of the segments of the diameter of the ellipse and hyperbola, terminated at the vertices; as a certain given line called the latus rectum, to the part of the diameter which lies between the vertices, and is called the latus transversum; so in non-parabolic curves of the second genus, the product of three ordinates, is to the product of the abscisses of the diameter between the ordinates and the three vertices of the curve, in a given ratio; in which ratio, if three lines are taken to three segments of the diameter between the vertices of the curve, each to each, then these three lines may be considered as the latera recta of the curve, and the three segments of the diameter between the vertices as its latera transversa. And as in the conic parabola, which has but one vertex to a diameter, the rectangle under the ordinates, is equal to the rectangle under the absciss between the vertex and ordinates, and a given straight line called the latus rectum: so in curves of the second genus, which have only two vertices to the same diameter, the product of the three ordinates is equal to the product of the two parts of the diameter cut off between the ordinates and the two vertices, and a certain given straight line, which may be called the latus rectum.

4. *Of the Ratio of the Products of the Segments of Parallel Lines.*

Finally, as in the conic sections, where two parallel chords are cut by two parallel chords, the first by the third, and the second by the fourth, the rectangle of the segments of the first is to the rectangle of the segments of the third, as the rectangle of the segments of the second, to the rectangle of the segments of the fourth : so also where four such straight lines occur in a curve of the second genus, each being cut in three points, the product of the segments of the first line will be to the product of the segments of the third as the product of the segments of the second to the product of the segments of the fourth.

5. *Of Hyperbolic and Parabolic Branches, and their Directions.*

All infinite branches of curves of the second and higher genus, like those of the first, are either of the hyperbolic or parabolic sort. I define a hyperbolic branch as one which constantly approaches some asymptote, a parabolic branch to be that which, although infinite, has no asymptote. These branches are easily distinguished by their tangents; for supposing the point of contact to be infinitely distant, the tangent of the hyperbolic branch will coincide with the asymptote, but the tangent of the parabolic branch being at an infinite distance, vanishes, and is not to be found. The asymptote to any branch is, therefore, found by seeking for the tangent to a point in that branch at an infinite distance. The direction of the branch may be found, by determining the position of a straight line parallel to the tangent referred to a point in the curve infinitely distant ; for such straight line will have the same direction as the infinite branch itself.

Section III.

The Reduction of all Curves of the Second Genus to four Cases of Equations.

All lines of the first, third, fifth, seventh, or odd orders, have at least two infinite branches extending in opposite directions; and all lines of the third order have two branches of the same kind, proceeding in opposite directions, towards which no other of their infinite branches proceed, (except only the Cartesian parabola).

Case I. (fig. 1.)

If the branches be hyperbolic, let GAS be their asymptote, and let CBc be any line drawn parallel to it, meeting the curve on each side (if possible); let this line be bisected in X, which will be the locus of a hyperbola, say $X\Phi$, one of whose asymptotes is AG. Let its other asymptote be AB, and the equation defining the relation between absciss AB and ordinate BC, if $AB = x$, $BC = y$, will be of the form always

$$xy^2 + ey = ax^3 + bx^2 + cx + d$$

where the terms b, c, d, a, e, designate given quantities, affected by their proper signs $+$ or $-$, of which any may be deficient, so that the figure, by reason of their absence, be not changed to a conic section. It may be, however, that this conic hyperbola coincides with its asymptotes; that is, the point X may fall in the straight line AB; and in that case the term $+ey$ is deficient.

Case II. (fig. 2.)

But if the straight line CBc is not bounded by the curve at each end, but only meets the curve in one point, draw AB any straight line given in position, meeting the asymptote AS in A, and draw another line BC parallel to that asymptote, and meeting the curve in C; then the equation expressing the relation of the ordinate BC and absciss AB always assumes the form

$$xy = ax^3 + bx^2 + cx + d$$

Case III. (*fig.* 3.)

But if the opposite branches are of the parabolic sort, let the straight line C B *c* be drawn, if possible, meeting each branch of the curve, and being bisected in B, the locus of B will be a straight line. Let this straight line be A B, terminating at any point A, then the equation expressing the relation of ordinate BC and absciss A B always assumes the form

$$y^2 = a x^3 + b x^2 + c x + d$$

Case IV. (*fig.* 4.)

But when C B *c* only meets the curve in one point, and, therefore, cannot be bounded at both ends, let that one point be C; and at the point B let C B *c* meet another straight line given in position, A B, and terminating at any point A, then the equation expressing the relation between ordinate BC and absciss A B always assumes the form

$$y = a x^3 + b x^2 + c x + d$$

The Names of the Curves.

In the enumeration of these cases of curves, we shall call that which is included within the angle of the asymptotes in like manner as the hyperbola of the cone, the *inscribed* hyperbola; that which cuts the asymptotes, and includes within its branches the parts of the asymptotes so cut off, the *circumscribed* hyperbola; that which, as to one branch, is inscribed, and, as to the other, circumscribed, we shall call the *ambigenous* hyperbola; that which has branches concave to each other, and proceeding towards the same direction, the *converging* hyperbola; that which has branches convex to each other, and proceeding towards contrary directions, the *diverging* hyperbola; that which has branches convex to contrary parts and infinite towards contrary sides, the *contrary branched* hyperbola; that which, with reference to its asymptote, is concave at the vertex, and has diverging branches, the *conchoïdal* hyperbola; that which cuts the asymptote in contrary flexures, having on both sides contrary branches, the *serpentine* hyperbola; that which intersects its conjugate, the *cruciform* hyperbola; that which intersects and returns in a loop upon itself, the *nodate* hyperbola; that which has two branches meeting at an angle of contact, and there stopping, the *cusped* hyperbola; that which has an infinitely small conjugate oval, *i. e.* a conjugate point, the *punctate* hyperbola; that which, from the impossibility of two roots, has neither oval, node, cusp, or conjugate point, the *pure* hyperbola.

In the same way we shall speak of parabolas, as converging, diverging, contrary branched, cruciform, nodate, cusped, punctate, and pure.

In the case of the first-mentioned equation, if ax^3 is positive, (fig. 5), the figure will be a triple hyperbola with six hyperbolic branches progressing to infinity alongside of three asymptotes, no one of which is parallel to another, two alongside of each hyperbola, on contrary sides. And these asymptotes, if the term bx^2 is not deficient, will cut one another at three points, making a triangle ($\mathrm{D}d\delta$); but if the term bx^2 is deficient, all the points will converge to one point. In the former case, take $\mathrm{AD} = \dfrac{-b}{2a}$ and $\mathrm{A}d = \mathrm{A}\delta = \dfrac{b}{2\sqrt{a}}$, join $\mathrm{D}d$, $\mathrm{D}\delta$; and $\mathrm{A}d$, $\mathrm{D}d$, $\mathrm{D}\delta$, will be the three asymptotes. In the latter case (fig. 6) draw any ordinate BC parallel to the principal ordinate AG, and in it produced in each direction take BF, $\mathrm{B}f$, equal to each other, and in the ratio to AB of $\sqrt{a} : 1$, join AF, $\mathrm{A}f$; and AG, AF, $\mathrm{A}f$, will be the three asymptotes. This hyperbola we call redundant, because it exceeds the conic hyperbola in the number of its hyperbolic branches.

In every redundant hyperbola, if neither the term ey be deficient nor $b^2 - 4ac$ be equal to $\pm\, 4ae\sqrt{a}$, the curve will have no diameter. If either of these cases occur, the curve will have a single diameter; if both, it will have three diameters. Now, the diameter always passes through the intersection of two asymptotes, and bisects all straight lines which are terminated by these asymptotes on both sides and are parallel to the third asymptote. And the absciss AB is a diameter of the curve wherever the term ey is deficient. I use the word diameter called absolute, here and hereafter, in its common acceptation; namely, as the absciss which has everywhere two equal ordinates to the same point, one on each side of it.

SECTION IV.

THE ENUMERATION OF CURVES.

Of the nine redundant Hyperbolas, having no Diameter, and three Asymptotes, making a Triangle.

WHEN a redundant hyperbola has no diameter (fig. 5), let the four roots be found of the equation $ax^4 + bx^3 + cx^2 + dx + \dfrac{e^2}{4} = 0$. Let these be AP, A$\varpi$, A$\pi$, A$p$; draw the ordinates PT, $\varpi\tau$, $\pi\gamma$, pt; then these will touch the curve in so many points T, τ, γ, t, and, by touching, give the limits of the curve, by which its species is determined.

For, if all the roots AP, Aϖ, Aπ, Ap (figs. 5, 7), are real and unequal, and of the same sign, the curve consists of three hyperbolas (inscribed, circumscribed, and ambigenous), and of an oval. One hyperbola lies towards D, another towards d, and the third towards δ; and the oval always lies within the triangle D$d\delta$ within the limits $\tau\gamma$, in which it is touched by the ordinates $\pi\gamma$ and $\varpi\tau$.

This is the 1st species.

If the two greatest roots Aπ, Ap (fig. 8), or the two least (AP, Aϖ, fig. 9), are equal to each other, and all of the same sign, the oval and circumscribed hyperbola will coalesce, their points of contact γ and t, or T and τ, coming together, and the branches of the hyperbola, intersecting one another, run on into the oval, making the figure nodate.

This is the 2d species.

If the three greatest roots Ap, Aπ, Aϖ (fig. 10), or the three least roots Aπ, Aϖ, AP (fig. 11), are equal to each other, the node becomes a sharp cusp; because the two branches of the circumscribed hyperbola meet at an angle of contact, and extend no farther.

This is the 3d species.

If the two middle roots Aϖ, Aπ (fig. 12), are equal, the points of contact τ and γ coincide, and, therefore, the oval vanishes to a point; and the figure consists of three hyperbolas, inscribed, circumscribed, and ambigenous, with a conjugate point.

This is the 4th species.

If two roots are impossible and the other two unequal, and of the same sign, (they cannot be affected by contrary signs), three pure hyperbolas result, without oval, node, cusp, or conjugate point; and these hyperbolas will either lie at the sides of the triangle made by the asymptotes, or at its angles; thus forming either the 5th (figs. 12, 13) or the 6th species (figs. 14, 15).

If two roots are equal, and the other two, either impossible, (figs. 16, 18) or real, and with different signs from the equal roots, (figs. 17, 19), the figure will be cruciform, two of the hyperbolas intersecting one another, either at the vertex of the asymptotic triangle, or at its base, (figs. 16, 17).

These are the 7th and 8th species.

Lastly, if all the roots are impossible (fig. 20), or all real and unequal, (fig. 21), two being positive and two negative; then there will result two hyperbolas at the opposite angles of the two asymptotes, with a serpentine hyperbola round the third asymptote.

This is the 9th species.

The above are all the possible cases of roots. For if two roots are equal to each other, and the other two also equal, the curve will be a conic section with a straight line.

2. *Of the twelve Redundant Hyperbolas, having but one Diameter.*

If a redundant hyperbola has but one diameter, let its absciss be A B, and obtain the three roots or values of x in the equation $ax^3 + bx^2 + cx + d = 0$.

If all these roots are possible and of the same sign, the figure will consist of an oval lying within the asymptotic triangle, (fig. 22) and of three hyperbolas at its angles, viz. circumscribed at the angle D, and inscribed at the angles d and δ.

This is the 10th species.

If the two greater roots are equal, and the third of the same sign, the branches of the hyperbola lying towards D (fig. 23), will intersect one another in a node, on account of the contact of the oval.

This is the 11th species.

If the three roots are equal, the hyperbola will be cusped, without an oval (fig. 24).

This is the 12th species.

If the two lesser roots are equal, the third being of the same sign, the oval vanishes to a point (fig. 25).

This is the 13th species.

In these last four cases, the hyperbola which lies towards D includes its asymptotes; the other two lie within their asymptotes.

If two roots are impossible, three pure hyperbolas will result, without oval, intersection, or cusp. Of this case there are four examples; viz. if the circumscribed hyperbola lies towards D, (fig. 25); if the inscribed hyperbola lies towards D, (fig. 26); if the circumscribed hyperbola lies beneath the base $d\,\delta$ of the triangle $D\,d\,\delta$, (fig. 27); if the inscribed hyperbola lies under the same base, (fig. 28).

These are the 14th, 15th, 16th, and 17th species.

If two roots are equal, and the third differing in sign, the hyperbolas will be cruciform; viz. two of the three intersecting each other either at the vertex of the asymptotic triangle or at its base, (fig. 29, 30).

These are the 18th and 19th species.

If two roots are unequal, and of the same sign, the third being of a different sign, two hyperbolas will result in the opposite angles of the two asymptotes, with an intermediate conchoïd, which will either lie at the same side of its asymptote as the asymptotic triangle, (fig. 31), or on the contrary side, (fig. 32).

These are the 20th and 21st species.

3. *Of the two Redundant Hyperbolas, with three Diameters.*

The redundant hyperbola, which has three diameters, consists of three hyperbolas included within their asymptotes, either at the angles of the asymptotic triangle (fig. 33), or at its sides (fig. 34).

These are the 22d and 23d species.

4. *Of the nine Redundant Hyperbolas, with three Asymptotes converging to a common point.*

If three asymptotes intersect one another in a common point, the 5th and 6th species are changed to the 24th, (fig. 6) ; the 7th and 8th to the 25th, (fig. 35); the 9th to the 26th, (fig. 36), where the serpentine curve does not pass through the intersection of the asymptotes ; and to the 27th, where it passes through that point, (fig. 37), in which case the terms b and d are deficient ; and the intersection of the asymptotes is the centre of the figure equally distant from all its opposite parts. These four species have no diameter.

The 14th and 16th species are also changed to the 28th, (fig. 38) ; the 15th and 17th to the 29th, (fig. 39) ; the 18th and the 19th to the 30th, (fig. 40) ; the 20th and 21st to the 31st, (fig. 41). These species have one diameter.

Lastly, the 22d and 23d species are changed to the 32d species, which has three diameters, all passing through the intersection of the asymptotes, (fig. 42).

All these changes are readily understood by conceiving the asymptotic triangle diminished till it vanishes to a point.

5. *Of the six Defective Hyperbolas, without a Diameter.*

If, in the case 1 of the equations, the term ax^3 is negative, the curve will be a defective hyperbola with one asymptote, and only two infinite branches, extending by the side of that asymptote. This asymptote is the first and principal ordinate AG. If the term ey is not deficient, the figure will have no diameter ; if deficient, it will have one diameter only. In the former case the species will be as follows :—

When the roots of the equation

$$a\,x^4 = b\,x^3 + c\,x^2 + dx + \frac{e^2}{4}$$

namely $A\,\pi$, $A\,P$, $A\,p$, $A\,\varpi$, are all possible and unequal, the curve will be a serpentine hyperbola, surrounding its asymptote in contrary flexure, together with a conjugate oval, (fig. 43).

This is the 33d species.

If the two middle roots $A\,P$, $A\,p$ (fig. 44), are equal, the serpentine and oval become joined, intersecting each other so as to make a node.

This is the 34th species.

If the three roots are equal, the node becomes a sharp cusp at the vertex of the serpentine hyperbola, (fig. 45).

This is the 35th species.

If the two greater of three roots with the same sign, ($A\,p$, $A\,\varpi$, fig. 47,) are equal to each other, the oval vanishes into a point.

This is the 36th species.

If any two roots are imaginary, there will only remain a pure serpentine hyperbola, without oval, intersection, cusp, or conjugate point. If this serpentine hyperbola does not pass through A, (fig. 46), the species is the 37th; but if it passes through A, (which happens when the terms b and d are deficient), then that point A will be the centre of the curve, and will bisect all straight lines terminated at each end by the curve, (fig. 47).

This is the 38th species.

6. *Of the seven Hyperbolas, Defective, but having Diameters.*

In the case where the term $e\,y$ is deficient, and when on that account the curve has a diameter, if all the roots of the equation $a\,x^3 = b\,x^2 + cx + d$, viz., AT, $A\,t$, $A\,\tau$, (fig. 48), are real, unequal, and of the same sign, the curve will be a conchoïdal hyperbola, with an oval at its convexity.

This is the 39th species.

If two roots are unequal, and of the same sign, the third being of a contrary sign, the oval will lie on the concave side of the conchoïd, (fig. 49).

This is the 40th species.

If the two lesser roots A T, A t (fig. 50), are equal, the third A τ being of the same sign, the oval and conchoïd coalesce, intersecting each other in a node.

This is the 41st species.

If three roots are equal, the node becomes a cusp, and the curve is the cissoïd of the ancients, (fig. 51).

This is the 42d species.

If the two greater roots are equal, the third being of the same sign, the conchoïd will have a conjugate point on its convex side, (fig. 52).

This is the 43d species.

If two roots being equal, the third have a contrary sign, the conchoïd will have a conjugate point on its concave side, (fig. 53).

This is the 44th species.

If two roots are impossible, a pure conchoïd results, without oval, node, cusp, or conjugate point, (figs. 52, 53).

This is the 45th species.

7. *Of the seven Parabolic Hyperbolas, having no Diameter.*

Whenever, in the first cited case of equations, the term $a x^3$ is deficient, the term $b x^2$ being present, the curve will be a parabolic hyperbola with two hyperbolic branches to one asymptote S A G, and two parabolic branches converging towards one and the same side. If the term $e y$ be present, the curve will have no diameter; if it be deficient, then it will have one diameter.

In the former case the following species will result:—

When three roots A P, A ϖ, A π (fig. 54), of the equation

$$b x^3 + c x^2 + d x + \frac{e^2}{4} = 0$$

are unequal, and of the same sign, the curve will consist of an oval and two other curves, part hyperbolic and part parabolic : that is, the parabolic branches will run into the hyperbolic branches next to them.

This is the 46th species.

If the two lesser roots are equal, the third being of the same sign, the oval and one of the two hyperbolo-parabolic branches become joined, and intersect each other in a node (fig. 55).

This is the 47th species.

If three roots are equal, this node becomes a cusp, (fig. 56).
This is the 48th species.

If the two greater roots are equal, and the third of the same sign, the oval vanishes into a conjugate point, (fig. 57).
This is the 49th species.

If two roots are impossible, the two hyperbolo-parabolic curves will be pure, without oval, intersection, cusp, or conjugate point, (fig. 57, 58).
This is the 50th species.

If two roots are equal, the third of a contrary sign, the hyperbolo-parabolic curves will intersect one another cross-wise, (fig. 59).
This is the 51st species.

If two roots being unequal, and of the same sign, the third is of a contrary sign, the curve becomes a serpentine hyperbola about its asymptote, (fig. 60), with a conjugate parabola.
This is the 52d species.

8. *Of the four Parabolic Hyperbolas, having a Diameter.*

In the other case, (that is, where the term *ey* is deficient, and the figure has a diameter,) if two roots of the equation

$$b x^2 + c x + d = 0$$

are impossible, two hyperbolo-parabolic curves will result equally distant from each side of the diameter **A B**, (fig. 61).

This is the 53d species.

If in this equation two roots are real and equal, the hyperbolo-parabolic curves unite, intersecting one another crosswise, (fig. 62).

This is the 54th species.

If the roots are unequal and of the same sign, a conchoïdal hyperbola will result, with a parabola on the same side of the asymptote, (fig. 63).

This is the 55th species.

If the roots are of different signs, the conchoïdal hyperbola and the parabola will be on different sides of the asymptote, (fig. 64).

This is the 56th species.

9. *Of the four Hyperbolisms of the Hyperbola.*

Whenever, in the first cited case of the equations, both the terms $a x^3$ and $b x^2$ are deficient, the curve will be a hyperbolism of some conic section. I call it the hyperbolism of a curve, when the ordinate is found by applying the product of the ordinate of that curve, and a given straight line, to a common absciss. In this manner a straight line is turned into a conic hyperbola, and every conic section is turned into some curve, which I here call the hyperbolism of a conic section.

For the equation to the curves of which we are speaking, (that is to say, $x y^2 + e y = c x + d,$) gives

$$y = \frac{e \pm \sqrt{e^2 + 4 d x + 4 c x^2}}{2 x}$$

which is formed by applying the product of the ordinate of the conic section

$$\frac{e \pm \sqrt{e^2 + 4 d x + 4 c x^2}}{2 m}$$

and a given straight line m, to the common absciss of the curves x.

From whence it appears that the curve will be a hyperbolism of the hyperbola, ellipse, or parabola, according as the term cx is positive, negative, or zero.

The hyperbolism of the hyperbola has three asymptotes, of which one is the prime and principal ordinate Ad, (fig. 65); the two others are parallel to the absciss AB, equally distant from it, and on each side of it. In the principal ordinate Ad, take Ad, $A\delta$, on each side equal to the quantity $\sqrt{c}$; and through the points d and δ draw dg, $\delta\gamma$, asymptotes parallel to the absciss AB.

Where the term ey is not deficient, the curve has no diameter: in this case, if the two roots (AP, Ap, fig. 65) of the equation

$$cx^2 + dx + \frac{e^2}{4} = 0$$

are possible and unequal (they cannot be equal unless the curve is a conic section), the curve will consist of three hyperbolas, opposite each other, of which one lies within the parallel asymptotes, and the other two outside of it.

This is the 57th species.

When these two roots are impossible, two opposite hyperbolas result outside of the parallel asymptotes, and a hyperbolic serpentine curve inside of them. This curve is of a double species. For when the term d is not deficient, it has no centre, (fig. 66); but when the term d is deficient, the point A is the centre, (fig. 67).

These are respectively the 58th and 59th species.

But if the term ey is deficient, the curve will consist of three opposite hyperbolas, one of which lies within the parallel asymptotes, and the other two outside them, like the 57th species; and, besides, it has a diameter, which is the absciss AB, (fig. 68).

This is the 60th species.

10. *Of the three Hyperbolisms of the Ellipse.*

The hyperbolism of the ellipse is expressed by this equation

$$xy^2 + ey = cx + d$$

and has but one asymptote, namely, the principal ordinate Ad, (fig. 69). When the term ey is not deficient, the curve is a serpentine hyperbola without a diameter, and even without a centre, if the term d is not deficient.

This is the 61st species.

But if the term d is deficient, the figure has a centre, but no diameter, its centre being the point A, (fig. 70).

This is the 62d species.

If the term ey is deficient, and the term d not deficient, the curve is a conchoïd to the asymptote AG, (fig. 71), and has a diameter without a centre, its diameter being the absciss AB.

This is the 63d species.

11. *Of the two Hyperbolisms of the Parabola.*

A hyperbolism of the parabola is expressed by the equation

$$xy^2 + ey = d$$

and has two asymptotes, namely, the absciss AB and the prime or principal ordinate AG. The hyperbolas in this curve are two, not lying within the opposite angles of the asymptotes, but in the adjacent angles, on each side of the absciss AB, and either without a diameter, where the term ey is present, (fig. 72), or with a diameter where ey is deficient, (fig. 73).

These two are the 64th and 65th species.

12. *Of the Trident.*

In the second cited case of the equations, we had the equation

$$xy = ax^3 + bx^2 + cx + d$$

In this case the curve will have four infinite branches, of which two are hyperbolic about the asymptote AG, (fig. 74), extending on contrary sides, and two are parabolic, converging and making with the other two a sort of trident-shaped figure. And this is the very parabolic curve by means of which Des Cartes constructed an equation of six dimensions.

This is the 66th species.

13. *Of the five Diverging Parabolas.*

In the third cited case of the equations we had the equation

$$y^2 = ax^3 + bx^2 + cx + d$$

which indicates a parabola, whose branches are divergent with respect to each other, and extend to infinity towards contrary parts.   The absciss A B is its diameter, and the species are the five following.

If the roots of the equation, $ax^3 + bx^2 + cx + d = 0$, are all possible, and unequal, Aτ, A T, At, the curve will be a divergent bell-shaped parabola, with an oval at the vertex, (figs. 74, 75).

This is the 67th species.

If two of the roots are equal, the parabola becomes either nodate, by running into the oval, (fig. 77), or punctate, on account of the oval becoming infinitely small, (fig. 78).

These two are the 68th and 69th species.

If the three roots are equal, the parabola will be cusped at the vertex, (fig. 80), and this is the Neilian parabola, commonly called the semicubic.

This is the 70th species.

If two roots are impossible, a pure bell-shaped parabola will result, (figs. 78, 79).

This is the 71st species.

14. *Of the Cubic Parabola.*

In the fourth case cited of the equations, if the equation be

$$y = ax^3 + bx^2 + cx + d$$

a parabola is indicated, having branches in contrary directions, and commonly called the cubic parabola, (fig. 81).

Thus, altogether, the species are 72.

Section V.

The Generation of Curves by Shadows.

If the shadows of curves caused by a luminous point, be projected on an infinite plane, the shadows of conic sections will always be conic sections; those of curves of the second genus will always be curves of the second genus; those of the third genus will always be curves of the third genus; and so on *ad infinitum*.

And in the same manner as the circle, projecting its shadow, generates all the conic sections, so the five divergent parabolas, by their shadows, generate all other curves of the second genus. And thus some of the more simple curves of other genera might be found, which would form all curves of the same genus by the projection of their shadows on a plane.

Of Double Points in Curves.

We have remarked that curves of the second genus can be cut by a straight line in three points. Sometimes two of these points coincide; as in the case when the straight line passes through an infinitely small oval, or through the intersection of two parts of the curve cutting each other, or meeting in a cusp. And whenever all the straight lines, extending in the direction of any infinite branch, cut the curve in only one point (as occurs in the ordinates of the Cartesian parabola, and of the cubic parabola, as well as in the straight lines of the absciss of the hyperbolisms of the hyperbola, and in the parallel lines of the parabola), we must conceive that those straight lines pass through two other points in the curve at an infinite distance. Two intersections of this sort when they coincide, whether at a finite or at an infinite distance, we shall call a double point. Now the curves possessing a double point may be described by the help of the following theorems.

Section VI.

Of the Organic Description of Curves.

Theorem I.

If (fig. 82) two angles of given magnitude P A D, P B D, are made to rotate about the poles A, B, given in position, and their legs A P, B P, at their concourse P, move along a straight line, the concourse of the other two legs A D, B D, will describe a conic section passing through the poles A, B; except when the straight line in question passes through either pole A or B, and except when the angles B A D, A B D, vanish together; in which cases the point D will describe a straight line.

Theor. II.

If the legs A P, B P, at their point of concourse P, pass along a conic section running through either pole, A, the two other legs A D, B D, at their point of concourse D, (fig. 83), will describe a curve of the second genus, passing through the other pole B, and having a double point in the first pole A, through which the conic section passes; except when the angles B A D, A B D, vanish together; in which case the point D will describe another conic section passing through the pole A.

Theor. III.

But if the conic section, which the point P passes along, runs through neither of the poles, A, B, (fig. 84), the point P will describe a curve of the second genus, or one of the third genus, having a double point. This double point will be found at the point of concourse of the legs A B, B D, which describe the curve, when the angles B A P, A B P, vanish together. But the curve described will be of the second genus, if the angles B A D, A B D, vanish together, otherwise it will be of the third genus, and will have two other double points at the poles A, B.

To describe Conic Sections by five given Points.

A conic section is determined when five points in it are given, and by them may thus be described. Let the five points be A, B, C, D, E (fig. 85). Join any three of them, A, B, C, and of the triangle A B C let any two of the angles C A B, C B A, rotate about their vertices A, B, and when C, the intersection of the two legs A C, B C, is successively applied to the two other points D, E, let the intersection of the remaining legs A B, B A, take place at the points P, Q. Let the straight line P Q be drawn and infinitely extended, and let the moveable angles be so rotated as that the intersection of the legs A B, B A, may run along the line P Q; then C, the intersection of the remaining legs, will describe the conic section proposed by the first theorem.

The Description of Curves of the Second Genus having a Double Point, by means of Seven given Points.

All curves of the second genus, having a double point, are determined where seven points in them are given, and may be described through those points as follows, one of these points being the double point:—Let any seven points A B C D E F G (fig. 86) in the curve to be described, be given, A being a double point. Join A, and any two other of the points, say B and C, and of the triangle A B C, let the angle C A B revolve about its vertex A, as well as either of the remaining angles A B C about its vertex B. And when the point C, being the concourse of the legs A C, B C, is successively applied to the four remaining points D, E, F, G; let the concourse of the other legs A B, B A, fall on the four points P, Q, R, S. Let a conic section now be described passing through the four points, and the fifth A; and let the aforesaid angles C A B, C B A, so rotate, that the concourse of their legs A B, B A, shall run along the said conic section, then the concourse of the remaining legs A C, B C, will describe the curve proposed in the second theorem.

If, instead of the point C, the straight line B C is given in position, which touches the curve to be described in B, the lines A D, A P, will coincide; and, instead of the angle D A P, there will be a right line to be rotated about the pole A.

If the double point A be infinitely distant, the straight line must be drawn continually extending in the direction of that point, and be moved in parallel motion while the angle A B C revolves about the pole B.

These curves may also be described somewhat differently by means of Theorem III.; but it is enough to have explained the more simple method here given.

In like manner, curves of the third, fourth, and higher genera may be described; not, indeed, all curves, but so many of them as by reason of some convenient relation, may be drawn by the motion of their loci. But to describe any curve whatever of the second or superior genus conveniently, when it has no double point, must be considered a very difficult problem.

Section VII.

The Construction of Equations by the Description of Curves.

One of the uses of geometrical curves is, that by means of their intersections, the solution of problems may be effected. Let an equation of nine dimensions be proposed for construction.

$$x^9 * + bx^7 + cx^6 + dx^5 + ex^4 + fx^3 + gx^2 + hx + k = 0$$
$$+ m$$

Where b, c, d, &c. signify any given quantities whatever, with their signs $+$ and $-$. Let the equation to the cubic parabola $x^3 = y$, be assumed, and the first equation becomes, by the substitution of y for x^3,

$$y^3 + bxy^2 + cy^2 + dx^2y + exy + my + fx^3 + gx^2 + hx + k = 0$$

which is an equation to a curve of the second genus. Where m, or f, may be deficient, or may be assumed at pleasure. By the description of these curves, and their intersections, the roots of the equation to be constructed will be given. It will be enough to describe the cubic parabola once only.

If the equation to be constructed be reduced to one of seven dimensions by the deficiency of the two last terms hx and k, the other curve, by destroying m, will have a double point at the beginning of the absciss, and hence may easily be described by the above rules.

If the equation to be constructed, by the deficiency of the three last terms $gx^2 + hx + k$, be reduced to one of six dimensions, the other curve, by destroying f, becomes a conic section.

And if, by the deficiency of the six last terms, the equation be reduced to one of three dimensions, we come to Wallis's construction by means of the cubic parabola and straight line.

Equations may also be constructed by means of the hyperbolism of the parabola with a diameter. Thus, if an equation of nine dimensions wanting the last term, be to be constructed,

$$a + cx^2 + dx^3 + ex^4 + fx^5 + gx^6 + hx^7 + kx^8 + lx^9 = 0$$
$$+ m$$

Let the equation to the hyberbolism in question be assumed

$$x^2 y = 1$$

Then substituting y for $\dfrac{1}{x^2}$, the equation becomes

$$ay^3 + cy^2 + dxy^2 + ey + fxy + m + g + hx + kx^2 + lx^3 = 0$$

which indicates a curve of the second genus, by whose description the problem will be solved. Here either the quantity m or g may be deficient or assumed at pleasure.

By means of the cubic parabola and curves of the third genus, equations may also be constructed of all dimensions up to twelve, and by the same parabola and curves of the fourth genus, equations may be constructed up to fifteen dimensions, and so on *ad infinitum*. These curves of the third, fourth, and higher genera, may always be found by means of description through points ascertained by geometrical means. Thus, if the equation to be constructed be

$$x^{12} * \; + ax^{10} + bx^9 + cx^8 + dx^7 + ex^6 + fx^5 + gx^4 + hx^3 + ix^2 + kx + l = 0$$

and a cubic parabola be described, let the equation to that parabola be $x^3 = y$, then substituting y for x^3, the equation to be constructed becomes

$$\begin{aligned}
y^4 &+ axy^3 + cx^2y^2 + fx^2y + ix^2 = 0 \\
&+ b \quad\;\; + dx \quad\; + gx \quad\; + hx \\
&+ e \qquad\quad\; + h \quad\;\; + l
\end{aligned}$$

which is an equation to a curve of the third genus, by the description of which the problem may be solved. This curve may be described by finding its points by means of plane geometry, because the indeterminate quantity x does not rise to higher dimensions than two.

London :—Printed by G. BARCLAY, Castle St. Leicester Sq.

NOTES TO THE FOREGOING TEXT.

In the following notes the undermentioned propositions are taken as proved : —

1. In any algebraic curve, if one of the co-ordinates becomes equal to zero, so that one variable in the equation which defines the curve vanishes, the roots of the equation so reduced are the values of the corresponding co-ordinate.

2. Infinite branches can only occur in pairs, the number of such branches is therefore always even.

3. If an algebraic curve has no infinite branch, it is necessarily an oval, that is, it returns upon itself.

4. The general indeterminate equation of the nth degree includes all the lines of the nth order.

5. If n denote the dimensions of a curve, $\dfrac{n^2 + 3n}{2}$ will be the number of constants in the general equation, defining all the curves of the order n.

Thus a straight line is determined by two points, its order being 1 ; a conic section by five points, its order being 2 : a cubic line by nine points, its order being 3, &c.

6. Every algebraic curve may have as many asymptotes as there are dimensions in the equation expressing it.

33

Notes to Section I.

On the Orders of Lines.

Newton here makes use of the word 'geometrical' in a limited Note to p. 7, line 5. sense in accordance with the meaning assigned to it by the mathematicians following Descartes, viz. as distinguishing lines which can be always indicated by an algebraical equation from those of which the equations are transcendental. There are many curves which can be cut by a straight line in only two points, but only one class of geometrical curves which can be so cut, namely, the conic sections. Curves, such as the catenary, &c., which admit of a straight line cutting them in only two points, but which cannot be expressed by an algebraical equation in finite terms, were formerly called mechanical curves; but this distinction of geometrical and mechanical, originally proposed by Descartes, has fallen into disuse, the division of curves into algebraical and transcendental being considered more appropriate.

Neither has the distinction adopted by Newton between lines and curves been generally followed by other writers, it being found more convenient to treat the straight line as a curve whose radius of curvature is infinitely great, so that the classification of curve lines may proceed according to the dimensions of the equation expressing the relation of their co-ordinates. A curve of the first order will thus be indicated by a simple equation between two variables, a curve of the second order by a quadratic, a curve of the third order by a cubic equation, and so on for the higher orders.

" . . *according to the number of points in which it can be cut by a* Note to p. 7, line 7. *straight line.*"

A line of the nth order is expressed by an equation of n dimensions which can have no more than n roots. A straight line therefore cannot cut a line of the nth order in more than n points for let the general equation representing a line of the nth order be assumed, and let the equation to a straight line referred to the same axes be

$$y = ax + b$$

then at the points where the straight line intersects the curve, the co-ordinates of the line and curve will be the same. Substituting the value of (y), viz. $a\,x + b$, in the general equation we have a resulting equation of the same dimensions, and therefore having the same number of roots as before, and these roots when real represent the abscissas of the points of intersection. If some of the roots are imaginary, there will be fewer points of intersection.

Note to p. 7, line 17.

"A line of the infinitesimal order is one which a straight line may cut in an infinite number of points, such as the spiral, cycloïd, quadratrix."

It is obvious that the spiral and the cycloïd (if we imagine its generating circle to be perpetually rolling on) are curves which may be cut by a straight line in an infinite number of points. It is not so immediately obvious that the quadratrix of the circle, which it is to be presumed is the curve alluded to by Newton, can be so cut. Indeed this curve, as it was known to the ancients, cannot be cut by a straight line in more than two points, because its inventor and those who followed him did not contemplate the continuance of the curve beyond the limits of its generating semicircle. It was first shown by Vincent Leotaud, a Jesuit professor of mathematics in the college of Dôle, about the year 1650, that by a more general consideration of the genesis of the curve it may become susceptible of two infinite branches, extending below the axis of x, and limited by asymptotes parallel to the axis of y, at distances $-a$, and $3\,a$ from the origin, (a) being the radius of the circle used. (See his "*Cyclomathia seu de Multiplici Circuli Contemplatione*," lib. iii.) If we imagine the rotation of the radius beyond 360°, the branches of the quadratrix become infinite in number as well as in extent: the curve is therefore one of infinitesimal order.

There is another quadratrix of the circle, invented by Tschirnhausen in the seventeenth century, which has no infinite branches, but still is a curve of infinitesimal order.   Its equation is, (a) being the radius of the circle,

$$y = a \sin \frac{\pi x}{2a}.$$

From which we infer that the curve is composed of an infinite number of sinuosities, meeting the axis of x at intervals equal to the diameter of the circle; for when $x = 2\,n\,a$, $y = o$, whatever value may be assigned to n; and thus a straight line may cut the curve in an infinite number of points.

Notes to Section II.

" Of Curves of the Second Genus, their Ordinates, Diameters, &c."

The proposition regarding the rectilinear diameters of lines of _{Note to p. 8, line 18.} the third order is, that if two parallel chords be drawn and be cut by a straight line, so that their sum on one side of it be equal to their sum on the other, then all lines drawn to meet the curve parallel to these will be cut by the straight line in the same manner.   The proposition may be shown to be true for all algebraic curves thus, let

$$y^n - (ax+b)\,y^{n-1} + (cx^2+dx+e)\,y^{n-2} - \&c. = o$$

Then, since in every equation the coefficient of the second term with its sign changed is equal to the excess by which the sum of the positive exceeds the sum of the negative roots, where this term is deficient, it is an indication that the sum of the positive is equal to the sum of the negative roots, that is, in the curve represented by the equation, that the sum of the positive equals the sum of the negative ordinates.

Let x and x' represent any two values of the absciss corresponding to intersections with two parallel ordinates, then the equation in the two cases becomes

$$y^n - (ax+b)\,y^{n-1} + \ldots \&c. = o, \text{ and } y^n - (ax'+b)\,y^{n-1} + \ldots \&c. = o$$

And since by hypothesis, in each case the sum of the positive equals the sum of the negative ordinates,

$$a x + b = 0, \text{ and } a x' + b = 0 \ \therefore \ a\,(x - x') = 0.$$

Hence a and b also $= 0$, showing that whatever value be assigned to the absciss, the coefficient of the second term in the equation vanishes; in other words, the sum of the ordinates on each side of the rectilinear diameter is equal for all values of x.

" The Intersection of two Diameters is the Centre."

Note to p. 8, line 30.

This definition is open to objection, for there may be a centre to a curve which admits of no rectilinear diameter. For instance, the curve No. 38 of the enumeration, fig. 47, and some of the curves whose asymptotes intersect in a point, as fig. 37. A curve may be said to have a centre when every value of a radius vector has its equal and opposite value. In the case of curves of uneven degree, this can only take place when the equation referred to the centre contains no term of even dimensions in x and y, the variables; in the case of curves of even degree the central equation can contain no term of uneven dimensions in x and y. To ascertain, therefore, whether a curve admits of a centre, it is sufficient to see whether the terms of odd (or even) dimensions of the variables can be exterminated by transforming the equation.

The conditions of a centre existing, and its position, may be readily obtained by differentiation; for instance, in the circle whose equation is

$$y^2 = 2\,a x - x^2$$

If this equation be differentiated and the terms made separately $= 0$, the result gives $x = a$, $y = 0$, the co-ordinates of the centre. In the case of a line of the third order, if the equation be twice differentiated, and the resulting terms made separately $= 0$, the values obtained for x and y will indicate the conditions which determine the existence of a general centre. Thus, in the first case of the Newtonian equations,

$$x y^2 + e y = a x^3 + b x^2 + c x + d$$

From the second differentiation we obtain

$$(1)\ 2x = 0, \quad (2)\ 4y = 0, \quad (3)\ (6\,ax + 2\,b) = 0 \ \therefore b = 0.$$

And since $d = 0$, also, the existence of a general centre depends on the conditions $b = 0$, $d = 0$, and the centre must be the origin itself.

" The sum of those two segments of the secant which are drawn from any two asymptotes on the same side to two points of the curve, will be equal to the third part which is drawn from the third asymptote on the contrary side to the third point of the curve."

Note to page 9, line 8.

In the figure, let AP be the diameter to the ordinates PM, PM', PM''; therefore, $PM + PM' = PM''$.

Now at an infinite distance the curve coincides with its asymptote; therefore the points M and H, M' and H', M'' and H'' coincide, when the secant is supposed at an infinite distance; so that $PH' + PH = PH''$, which equation will be true for any position of the secant parallel to itself;

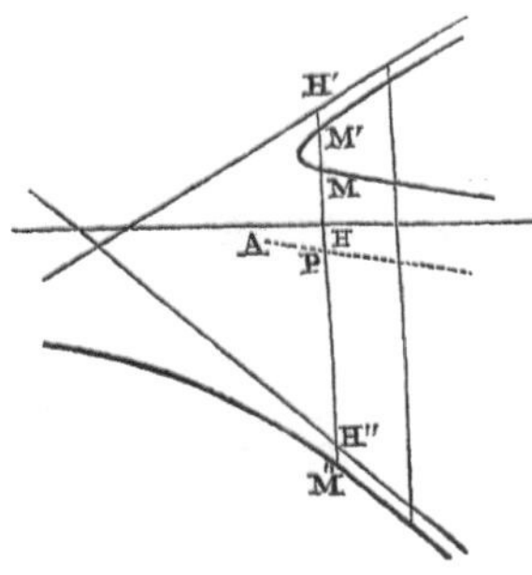

$$\therefore PM - PH + PM' - PH' = PM'' - PH'', \text{ and}$$

$$HM - H'M' = H''M'', \text{ or}$$

$$HM = H'M' + H''M''.$$

Where there are three rectilinear asymptotes, the three branches of a curve cannot be all situated on the same side relatively to such asymptotes; but if two branches lie beneath two asymptotes, the third lies above it. For instance, such a case as is represented in the figure is impossible; for the segments of the secant being all affected by the same sign, their sum will not be equal to 0, as it should.

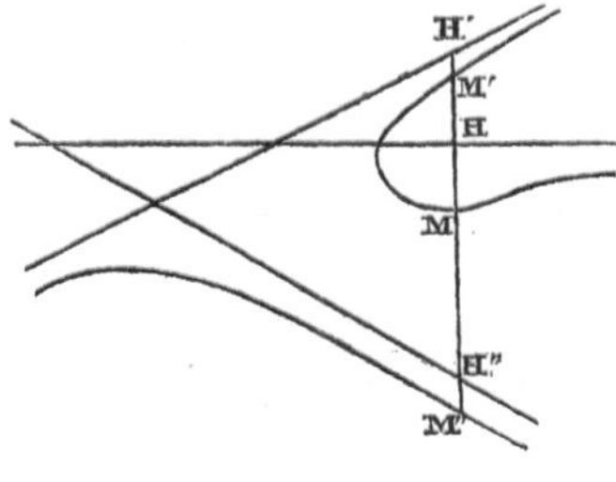

Note to
page 9,
line 20.

" *In non-parabolic curves of the second genus, the product of three ordinates is to the product of the abscissas of the diameter between the ordinates and the three vertices of the curve, in a given ratio.*"

Let the absciss A M and ordinate M P each cut the curve in three points; the product of the ordinates, viz., $M\pi . M\varpi . MP$ will be to the product of the abscissas $MK . MI . MH$ in an invariable ratio; the inclination of the axes being given.

In the general equation representing all lines of the third order, the last term, or that in which no power of y is involved, viz.,

$$f x^3 - g x^2 + h x - k,$$

is, by the theory of equations, equal to the product of all the roots. Therefore,

$$M\pi . M\varpi . MP = f x^3 - g x^2 + h x - k.$$

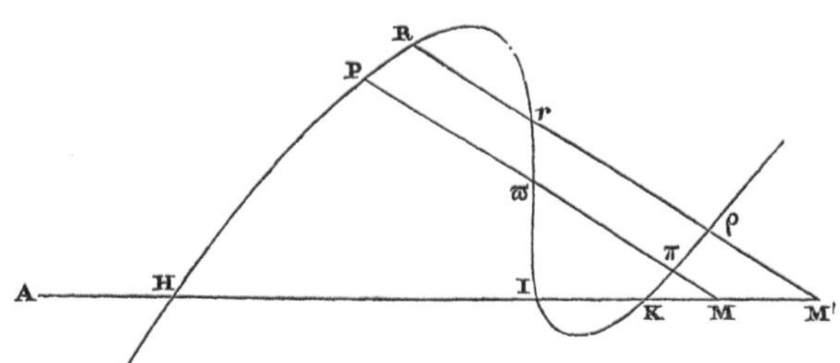

Let the three points in which the absciss cuts the curve be H, I, K: then A being the origin, A H, A I, A K, will be the values of x when $y = 0$, and will represent the three roots of the equation,

$$x^3 - \frac{g x^2}{f} + \frac{h x}{f} - \frac{k}{f} = 0.$$

This equation is formed by the three factors, $x - AH$, $x - AI$, $x - AK$, multiplied together; that is,

$$MH . MI . MK = x^3 - \frac{g x^2}{f} + \frac{h x}{f} - \frac{k}{f} = \frac{1}{f}(M\pi . M\varpi . MP);$$

or, the product of the ordinates between M and the curve is to the product of the abscisses between M and the curve in the invariable ratio $f : 1$.

A consideration of the general equation, without referring to geometrical ideas, leads to the same result, for in the equation

$$ay^3 + bxy^2 + cx^2y + dx^3 + ey^2 + fxy + gx^2 + hx + iy + k = 0$$

if x and y are successively assumed $= 0$

$$ay^3 + ey^2 + iy + k = 0 \qquad (1)$$
$$dx^3 + gx^2 + hx + k = 0 \qquad (2)$$

The product of the roots of (1) is $\dfrac{k}{a}$

and the product of the roots of (2) is $\dfrac{k}{d}$

They are therefore to each other in the invariable ratio $d : a$, which ratio will not be affected by any change of origin, but will therefore remain the same for all axes of co-ordinates parallel to the axes assumed.

We have only had occasion to notice the case in which both absciss and ordinate meet the curve in three points: but suppose the absciss cuts the curve in only one point, and let that point be the origin, we have

$$\mathrm{M}\pi \cdot \mathrm{M}\varpi \cdot \mathrm{MP} = fx^3 - gx^2 + hx = fx\left\{ \left(x - \frac{g}{2f} \right)^2 + \left(\frac{h}{f} - \frac{g^2}{4f^2} \right) \right\}$$

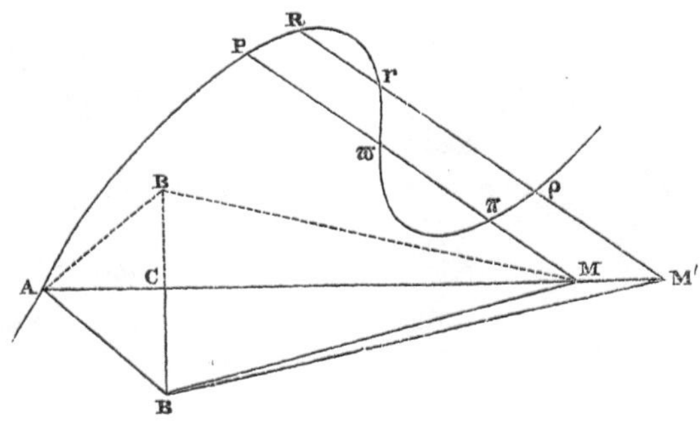

Take AC towards $\mathrm{M} = \dfrac{g}{2f}$, and draw CB perpendicular to AM and $= \dfrac{\sqrt{4fh - g^2}}{2f}$, then since $\left(x - \dfrac{g}{2f} \right)^2 = \mathrm{CM^2}$, and

$$\frac{h}{f} - \frac{g^2}{4f^2} = \mathrm{CB^2}$$

$$f \cdot \text{AM} \, (\text{CM}^2 + \text{CB}^2) \text{ or } f \cdot \text{AM} \cdot \text{BM}^2 = \text{MP} \cdot \text{M}\varpi \cdot \text{M}\pi.$$

Therefore when the absciss meets a line of the third order in only one point, the product of the ordinates is to the product of the absciss multiplied by the square of a given line B M, in an invariable ratio. That the perpendicular B C will always have a real value, appears from the consideration that whenever the absciss A M cuts the curve only once, the roots of the quadratic

$$f x^2 - g x + h = 0$$

are necessarily imaginary, and therefore $4 f h > g^2$, so that $\sqrt{4 f h - g^2}$ is real.

Even when the absciss does not meet the curve anywhere, in which case the equation for determining the abscisses is

$$g x^2 - h x + k = 0, \ h^2 \text{ being } < 4 g k,$$

the theorem still holds good, for a line may always be found such that its square shall be to the product of the ordinates in an invariable ratio. Thus let the absciss A M to the ordinate M P be assumed so as not to meet the curve.

$$\text{M}\pi \cdot \text{M}\varpi \cdot \text{MP} = g x^2 - h x + k = g \left\{ \left(x - \frac{h}{2g} \right)^2 + \frac{4 g k - h^2}{4 g^2} \right\}$$

Take $\text{A B} = -\dfrac{h}{2g}$, and draw the perpendicular B C H. With centre A, and radius $\sqrt{\dfrac{k}{g}}$, describe a circle determining the point

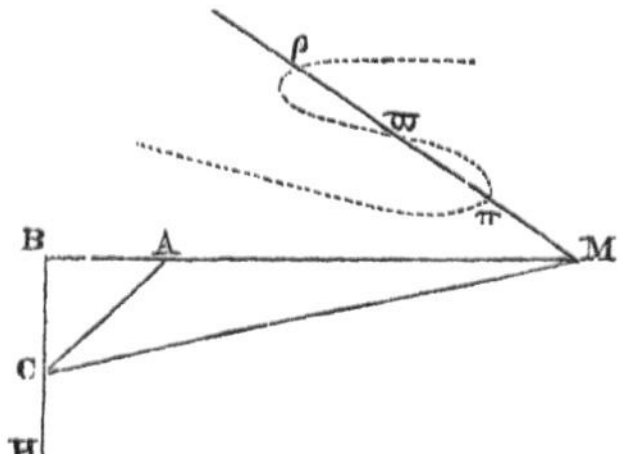

C, then C M will be the required line. It is evident that the point C is fixed, whatever may be the position of the point M in the absciss A M.

See Cramer, Analyse des Lignes Courbes, ch. v., Maclaurin, Lin. Geom. and Stirling's Commentary.

" Of the Ratio of the Products of the Segments of Parallel Lines."

We have seen that where the absciss and ordinate each cut the curve in three points, the product of the three ordinates is to the product of the three abscisses in a given ratio. In the curve A B C D let the absciss and ordinate cut the curve in the points

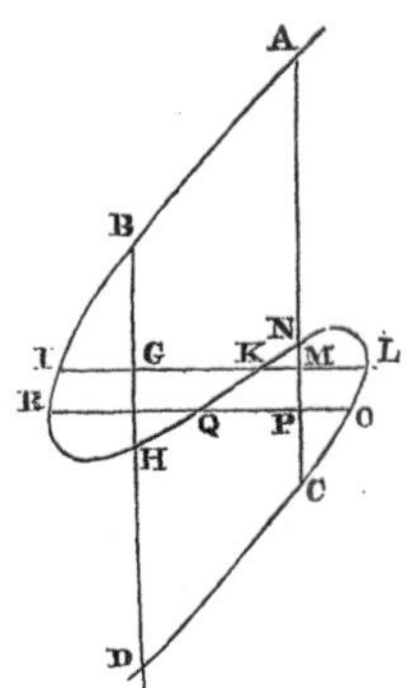

IKLBHD, and draw the parallel ordinate ANC.   The product

$$GB \cdot GH \cdot GD : GI \cdot GK \cdot GL ::$$
$$MA \cdot MN \cdot MC : MI \cdot MK \cdot ML$$

and drawing RO parallel to IL, we have for the same reason

$$MA \cdot MN \cdot MC : MI \cdot MK \cdot ML ::$$
$$PA \cdot PN \cdot PC : PR \cdot PQ \cdot PO$$

therefore

$$GB \cdot GH \cdot GD : GI : GK \cdot GL ::$$
$$PA \cdot PN \cdot PC : PR \cdot PQ \cdot PO$$

which is the proposition in the text.

"I define a hyperbolic branch as one which constantly approaches some asymptote, a parabolic branch to be that which, although infinite, has no asymptote." Note to page 10, line 15.

Here it is to be understood that rectilinear asymptotes are spoken of, and the definition suggests a simple method of distinguishing hyperbolic from parabolic branches, for if we express the ordinate to an infinite branch in the form of a series of descending powers of the absciss, and take all the terms of this series in which the exponent of x is positive or zero, then if the sum of these terms indicates a straight line, this straight line is an asymptote to the curve, which is therefore hyperbolic.   If the sum of the above-mentioned terms does not indicate a straight line the curve is parabolic.

Thus in the equation

$$y^3 + axy - x^3 = 0$$

the series for y will be

$$y = x - \frac{a}{3} + \frac{a^3}{3^4 x^2} \ \&c.$$

where the sum of the terms of the series in which the exponent of x is positive or zero, viz. $x - \dfrac{a}{3}$, indicates the ordinate of a straight line, and the curve is therefore hyperbolic as to its branch.

So in the equation

$$y^2 - x^3 + ax^2 = 0$$

the series for y will be

$$y = x^{\frac{3}{2}} + \&c.$$

which is not indicative of a straight line, and therefore the branch will be parabolic.

If the series for y *commence* with a negative power of x, as

$$y = \frac{a^2}{x^2} - \&c.$$

the branch of the curve will be hyperbolic, for when x is supposed infinitely great $\dfrac{a^2}{x^2}$ becomes equivalent to o, and the branch coincides with the straight line $y = $ o, that is, the absciss is itself an asymptote.

Since several different series may be derived from one equation, and these series may commence with different terms, the same curve may have both hyperbolic and parabolic branches.

Note to
page 10,
line 23.

" The asymptote to any branch is therefore to be found by seeking for the tangent to a point in that branch at an infinite distance."

This is the principle on which is founded the method usually given in modern treatises, for drawing the asymptotes to a curve whose equation is given; for the limiting position of the tangent, when the point of contact is infinitely remote, becomes asymptotic: and if its position is unlimited there will be no rectilinear asymptote.

There are, however, sometimes more convenient methods for determining the existence of the asymptotes of curve lines than this. The following is the method proposed by Stirling in his " Lineæ tertii ordinis Newtonianæ."

Let the ordinate be reduced to the form

$$y = A x^n + B x^{n-r} + C x^{n-2r} + \&c.,$$

converging for large values of x.

Assume a new ordinate, y', equal to all the initial terms of this series which are not diminished by the increase of x. Then the difference of these two ordinates as x increases must continually diminish, and ultimately vanish; therefore the branches having the ordinates y and y' to the same absciss x, will continually tend to coincidence, and y' is the ordinate to the asymptote, whose equation is therefore given.

This method enables us to perceive not only the nature of the asymptote, whether rectilinear or curvilinear, but also on which side of the curve it lies; for if the first term in the series which is diminished by the increase of x, that is the first term which contains the inverse power of x, is positive, the asymptote lies between the curve and absciss, if not, then the asymptote lies on the other side of the curve from the absciss.

And it will not be necessary in every case investigated, to have recurrence to this particular series. For we may assume one general series, applicable to all the curves included in the general equation belonging to any order, and then we may construct a general canon which will be sufficient for the investigation of the asymptotes of all such curves. Thus let the general equation of the second order be

$$A y^2 + B x y + C x^2 + D y + E x + F = 0.$$

Let $y = ax + b + cx^{-1} + \&c.$

Determining coefficients, we shall have $a = $ a root of the equation

$$A a^2 + B a + C = 0, \text{ whence } a \text{ is given}$$

$$b = -\frac{D a + E}{2 a A + B}, \quad c = \frac{A b^2 + D b + F}{2 a A + B}$$

Again, if we assume the general equation of the 3d order,

$$A y^3 + B x y^2 + C x^2 y + D x^3 + E y^2 + F x y + G x^2 + H y + K x + L = 0$$

which includes all lines of the 3d order whatever, then a will be a root of the equation

$$A a^3 + B a^2 + C a + D = 0$$

$$b = -\frac{A a^2 + B a + C}{3 E a^2 + 2 F a + C}$$

$$c = -\frac{3 A a b^2 + B b^2 + E a b + F b + H a + K}{3 A a^2 + 2 B a + C}$$

From which expressions may be found the asymptotes of all these curves without further recourse to series. For a always gives the inclination of the asymptote to the absciss; b gives the distance between the origin and intersection of the asymptote with the absciss; and c shows on which side of the asymptote lie the branches of the curve.

The asymptotes of a curve may also be found by means of Newton's parallelogram, of which examples will be given in a subsequent part of this volume.

Asymptotes of lines of an order superior to the conic sections may be cut by the curve, and the number of points in which they may so intersect the curve will be less by two than the dimensions of the equation of the curve. This may be shown as follows:—

Let the equation to the curve be

$$m y^n + (a x + b) y^{n-1} + (c x^2 + d x + e) y^{n-2} + \text{ &c. } = 0$$

If the curve has an asymptote parallel to the axis of y, the term $m y^n$ will vanish, since $m = 0$ in this case, and the x to which this asymptotic ordinate corresponds, will be determined by the equation $a x + b = 0$, or $x = -\dfrac{b}{a}$. Substituting which value in the equation to the curve, the term $(a x + b) y^{n-1}$ will also vanish, and then the term

$$\frac{c b^2 - d b a + e a^2}{a^2} y^{n-2}$$

becomes the first term in the equation. Therefore for the absciss $-\dfrac{b}{a}$, the equation whose roots determine the points where the asymptote and curve intersect, will be of $n - 2$ dimensions, and will not have more than $n - 2$ roots. The asymptote and curve can therefore intersect in $n - 2$ points only.

Cor. Thus a line of the third order can intersect its asymptote only once, *i.e.* each of the asymptotes may be once intersected by a branch, but only once.

This theorem admits of a simple geometrical illustration, for let A B C represent a line of the third order, which the straight line G B D cuts in three points. If the straight line be made to revolve on the pole B until it becomes parallel to the asymptote

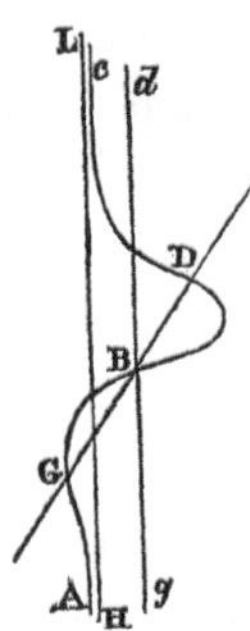

H L, it evidently will intersect the curve in only two points, the point G going off to infinity. And if the straight line continuing parallel to the asymptote be now conceived to move towards the asymptote H L, until coincidence takes place, the point of intersection D will also go off to infinity, and there will remain only B, in which the curve cuts the asymptote.

From the property of the asymptote here adverted to, it follows that if we draw the ordinates to the curve parallel to it, they will cut the curve in one point less than the dimensions of the equation defining it.

Thus an ordinate parallel to an asymptote of a line of the third order will only rise to quadratic dimensions, so that by a simple transformation we are enabled to describe any species whose branches are hyperbolic with facility, since to find the values of the ordinate to any point in the absciss, only the solution of a quadratic is needed.

Notes to Section III.

" *All lines of the 1st, 3rd, 5th, 7th, or odd orders, have at least two infinite branches extending in opposite directions.*"

Note to page 11, line 4.

Since impossible roots can only enter equations by pairs, it follows that every equation of an odd order, and therefore having an odd number of roots, must have one of those roots real. Therefore, however great the absciss belonging to such a curve be assumed, it must always have a real ordinate indicating a point in the curve, whose branch is consequently infinite. Also since the absciss may be assumed negative as well as positive, there will be a corresponding infinite branch on the negative side, and the curve has therefore at least two infinite branches.

One pair of infinite branches, then, is the least number which a line of uneven order must necessarily have, but it may have more than one pair, in fact as many pairs of branches as there are unities in the dimensions of the equation.

Lines of the second, fourth, sixth, and even orders, on the

other hand, will frequently have no infinite branches, the ordinates becoming impossible for any value of the absciss, positive or negative. Take, for example, the equation

$$y^4 + 2\,y^2 x^2 + x^4 - 6\,a x y^2 - 2\,a x^3 + a^2 x^2 = 0$$

whence

$$y = \pm \sqrt{ax} \pm \sqrt{2\,ax - x^2}$$

the four values of y, corresponding to any given absciss of a line of the fourth order, are here shown, and are evidently made up of the sums and differences of the ordinates of the parabola $y = \pm \sqrt{ax}$, and of the circle $y = \pm \sqrt{2\,ax - x^2}$, referred to the same axis and origin.

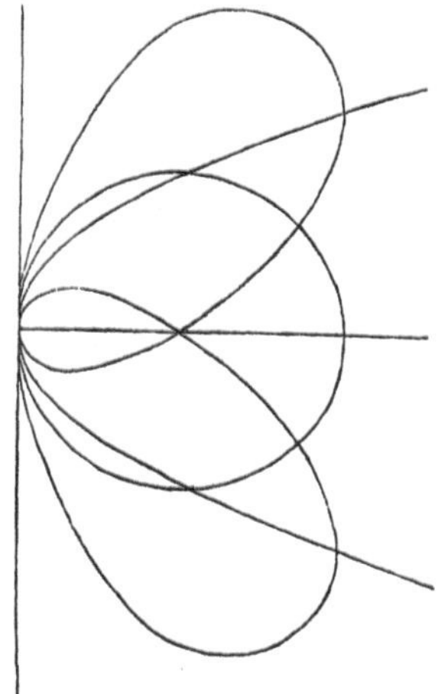

Now since the circle cannot extend beyond its diameter, it is evident that the curve cannot extend beyond the extremity of the circle's diameter, and must therefore return upon itself; consequently it has no infinite branches: the annexed figure represents the course of the curve.

Again the equation

$$2y = \pm \sqrt{6x - x^2} \pm \sqrt{6x + x^2} \pm \sqrt{36 - x^2}$$

represents a line of the eighth order, having four positive and four negative ordinates to each absciss. And it is evident if x is assumed either to be negative, or greater than 6, that then the ordinates will be imaginary, and that there are no infinite branches. The curve may be easily described by points, for we have the values of y

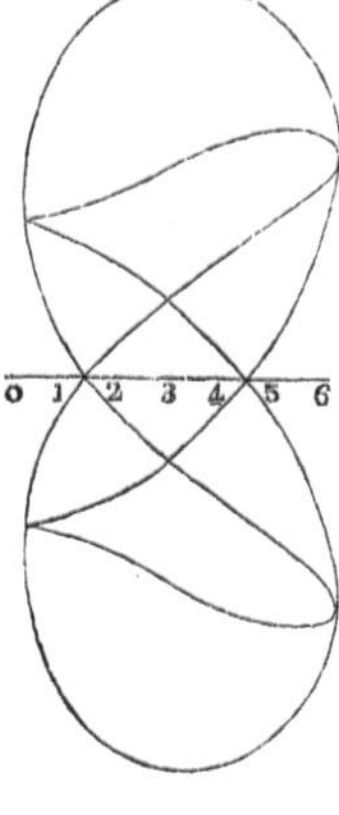

When	$x = 0$	$x = 1$	$x = 2$	$x = 3$	$x = 4$	$x = 5$	$x = 6$
$+ + + y =$	3·00	5·40	6·25	6·70	6·81	6·48	4·24
$- + + y =$	3·00	3·16	3·41	3·70	3·98	4·25	4·24
$+ - + y =$	3·00	2·75	2·25	1·50	0·50	−0·93	−4·24
$+ + - y =$	−3·00	−0·51	0·59	1·50	2·35	3·16	4·24

The figure shows the course of the curve; the example is taken from Euler's *Analysis Infinitorum*.

Case I. of the Equations.

" If the branches be hyperbolic, let GAS be their asymptote," &c.   Note to page 11, line 10.

The reduction of all lines of the third order to four cases of equations, is the basis upon which Newton has founded the whole system of the subsequent enumeration; he has, however, given no demonstration of this fundamental principle of his classification of these curves.

To demonstrate conclusively that it is always possible to reduce the general equation of the third order to one of the four cases proposed, would be to exceed the limits of a note. The French mathematician Nicole, has taken the trouble to discuss all the equations, thirty in number, which result from assuming each of the terms and combinations of terms in the general equation to vanish; see " Mémoires de l'Academie," for the year 1729, p. 194. It will be enough here to state that he arrives at the same conclusions as those of Newton.

Assuming then that the reduction in question can always be effected, we may proceed to show that the asymptote GAS and any chord ordinate CN parallel to it being given, the point which bisects this chord ordinate is the locus of a conic section, or else a straight line.

When the general equation is referred to axes one of which is the given asymptote, any point in it, S being taken as the origin, it appears in the form

$$xy^2 + (bx^2 + cx + d)y = fx^3 + gx^2 + hx + k,$$

whence

$$y = \frac{bx^2 + cx + d}{2x} \pm \sqrt{\frac{fx^3 + gx^2 + hx + k}{x} + \left(\frac{bx^2 + cx + d}{4x}\right)^2}$$

or for brevity write $y = \mathrm{P} \pm \mathrm{Q}.$

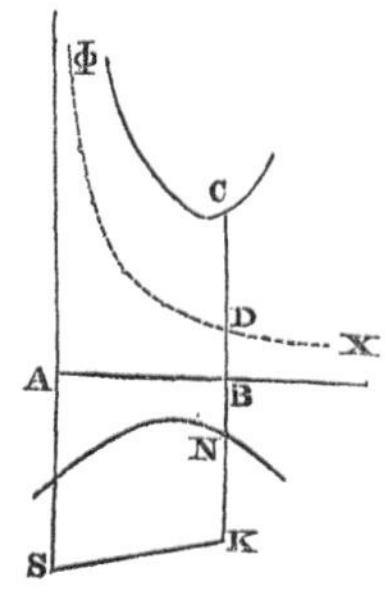

Let S K be the absciss to the ordinate K N C. Then $KC = P + Q, KN = P - Q, CN = 2Q.$ Bisect C N in D, and D will be the locus of a hyperbola, for $KD = P, (DC \& DN = \pm Q)$

that is, $y = \dfrac{bx^2 + cx + d}{2x}$, a hyperbola.

When $d = 0$, $y = \dfrac{bx + c}{2}$, an equation to a straight line, which explains the meaning of the author when he says, " It may be that this conic hyperbola coincides with its asymptotes, that is, the point X may fall in the straight line A B."

When the term x is not found in the denominator, and $y = \dfrac{bx^2 + cx + d}{2} \pm \&c.,$ the bisecting line is a parabola, and the curve is a parabolic hyperbola.

If the hyperbola $X\Phi$ be referred to its asymptotes as axes, its equation will appear in the form $y' (= BD) = \dfrac{e}{2x'}$, where $\dfrac{e}{2}$ is some constant quantity: and if the proposed curve be referred to the same axes, the sum of the positive and negative ordinates

$(BC \text{ and } BN) = 2BD.$ For $\begin{array}{l} BC = BD + DC \\ BN = BD - DC \end{array}.$

Therefore in the equation to the curve, $\dfrac{e}{x'}$, will be the co-efficient of y' in the transformed quadratic, and the form of it will be

$$y'^2 - \frac{ey'}{x'} = A x'^2 + B x' + C + \frac{D}{x'}$$

or,

$$x' y'^2 - ey' = A x'^3 + B x'^2 + C x' + D$$

which agrees with the first case of the equations given by Sir Isaac Newton.

Note to page 13, line 12.   *" In the case of the first-mentioned equation if $a x^3$ is positive (fig. 5) the curve will be a triple hyperbola with six hyperbolic* *branches progressing to infinity, alongside of 3 asymptotes no one of which is parallel to another."*

The proposed equation is

$$xy^2 - ey = ax^3 + bx^2 + cx + d$$

whence is obtained

$$y = \frac{e}{2x} \pm \sqrt{ax^2 + bx + c + \frac{d}{x} + \frac{e^2}{4x^2}}$$

Expanding the irrational portion of the equation in a series, we have

$$\pm \frac{e}{2x} \pm \frac{d}{e} \pm \&c.,$$

the two values of the ordinate will therefore be,

$$y = \frac{e}{x} + \frac{d}{e} + Ax + Bx^3 + \&c. \qquad (1)$$

$$y = -\frac{d}{e} - Ax - Bx^2 - \&c. \qquad (2)$$

Examining the curve in the neighbourhood of the origin, from (1) we infer that when $x = 0$, y is infinite and therefore the axis of y is an asymptote. From (2) we infer that the curve cuts this asymptote at a distance $-\dfrac{d}{e}$ from the axis of x.

This asymptote has two infinite branches of the curve alongside it, one on each side, in contrary directions.

If x is continued infinitely in a positive direction, there will always be a positive and a negative value for y, whence we infer the existence of two more infinite branches.

If x is negative the equation becomes

$$- xy^2 - ey = - ax^3 + bx^2 - cx + d,$$

or,

$$xy^2 + ey = ax^3 - bx^2 + cx - d.$$

Here also, when x is continued infinitely, there will be two values for y, one positive and the other negative. The number of infinite branches will therefore be, in all, six.

To show that no two of the asymptotes are parallel to one another, let y be reduced to a series converging for large values of x, there will result

$$y = x \sqrt{a} + \frac{b}{2 \sqrt{a}} + \frac{4\,ac - b^2 + 4\,ae \sqrt{a}}{8\,ax \sqrt{a}} + \&\text{c.}$$

$$y = - x \sqrt{a} - \frac{b}{2 \sqrt{a}} + \frac{4\,ac - b^2 - 4\,ae \sqrt{a}}{8\,ax \sqrt{a}} - \&\text{c.}$$

the first two terms of these series indicating the equations to two straight lines, viz.,

$$y = \pm \left(x \sqrt{a} + \frac{b}{2 \sqrt{a}} \right)$$

show the existence of two rectilinear asymptotes, cutting the axis of x at a distance $-\dfrac{b}{2\,a}$ from the origin, and making angles with this axis whose trigonometrical tangents are $\sqrt{a}$ and $-\sqrt{a}$.

Thus (in fig. 5) if $A\,D = \dfrac{b}{2\,a}$ and $A\,d$, $A\,\delta$, $= \dfrac{b}{2 \sqrt{a}}$, joining $D\,d$, $D\,\delta$, and producing $D\,d$, $D\,\delta$, $d\delta$, indefinitely, the three asymptotes of the six hyperbolic branches are apparent, and form a triangle. When the term $b\,x^2$ is deficient in the equation, $b = 0$, whence the term $\dfrac{b}{2 \sqrt{a}}$ vanishes, that is, the distance between the intersection of the asymptotes and the origin disappears, and the triangle becomes evanescent, the three asymptotes intersecting at a point. When this case occurs, the equation to the asymptotes will be represented by

$$y = x \sqrt{a}, \quad y = - x \sqrt{a}$$

or, as in the text, $B\,F$ or $B\,f : A\,B :: \sqrt{a} : 1.$

The geometrical signification of the series for y will be evident by an inspection of fig. 5.

$\pm x \sqrt{a}$ is that part of B C or B c which $= d \delta$.

$\pm \dfrac{b}{2a}$, is the part of B C or B c extending from the last-named point to the asymptote.

The remainder of the series consists of terms containing descending powers of x, and is represented by the intercept between asymptote and curve, so that when x is made infinite the third term of the series vanishes, the curve and its asymptote coinciding.

"*In every redundant hyperbola, if neither the term ey be deficient, nor $b^2 - 4ac$ be equal to $\pm 4ae \sqrt{a}$, the curve will have no diameter.*" Note to page 13, line 28.

The proposition put affirmatively is that all redundant hyperbolas have a diameter (1) when $e = 0$, (2) when $b^2 - 4ac = \pm 4ae\sqrt{a}$

Case I. when $e = 0$.

In this case

$$y = \pm \sqrt{ax^2 + bx + c + \frac{d}{x}}$$

The absciss here evidently bisects parallel chords, since the corresponding positive and negative ordinates are equal. There is therefore a diameter. When x is assumed infinitely small, $y = \dfrac{d^{\frac{1}{2}}}{x^{\frac{1}{2}}}$, and the exponent of x in the denominator being even, the infinite branches of the curve lie on the same side of the asymptote to which the ordinates are taken parallel, and therefore cannot intersect it.

It is a test of the existence of a diameter, that the curve does not intersect the asymptote which is assumed to be parallel to the bisected ordinates.

$$\text{Case II. when } b^2 - 4ac = \pm\, 4ae\, \sqrt{a}.$$

Here the ordinate to the asymptote is $y = \dfrac{b + 2ax}{2\sqrt{a}}$

and the ordinate to the curve is

$$y' = \frac{\frac{1}{2}e \pm \sqrt{ax^4 + bx^3 + cx^2 + dx + \dfrac{e^2}{4}}}{x}$$

And these ordinates are equal at the point where the curve and asymptote intersect, therefore when $y = y'$

$$\left(\frac{b^2}{4a} - e\sqrt{a} - c\right)x = d + \frac{be}{2\sqrt{a}}$$

or,

$$x = \frac{4ad + 2\sqrt{a}\,be}{b^2 - 4ac - 4ae\sqrt{a}}$$

is the value of the corresponding absciss. Here the condition $b^2 - 4ac = 4ae\sqrt{a}$ renders the denominator $= 0$, and x is therefore infinite, that is, the curve and asymptote have no point of intersection. There is, consequently, a diameter in this case bisecting ordinates parallel to the asymptote assumed.

It appears, then, that the redundant hyperbola admits of a diameter in the following three cases,

$$\text{when } e = 0$$

$$\text{when } e = \frac{4ac - b^2}{4a^{\frac{3}{2}}}$$

$$\text{when } e = \frac{b^2 - 4ac}{4a^{\frac{3}{2}}}$$

and it is evident that if the ordinates parallel to two of the asymptotes have a diameter, the ordinates parallel to the third asymptote will necessarily have one also, since of these three equations of condition, two cannot be true unless the third is also. There will therefore belong to the redundant hyperbolas, either one diameter, three diameters, or none.

In the diagram the lines A B, C D, E F, represent the three diameters of a redundant hyperbola bisecting ordinates parallel to the three asymptotes.

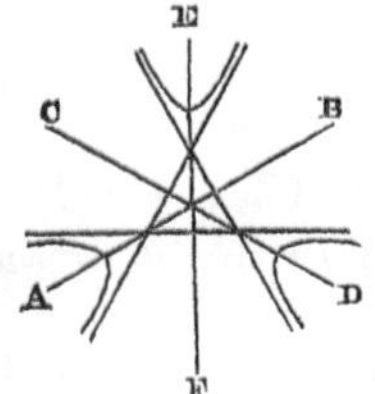

The curve which has but one diameter cuts the two asymptotes through the intersection of which the diameter passes, but not the third asymptote. The curve that has three diameters will not cut any of the asymptotes.

Diameters called ' absolute.'

It is evident from the definition here given of an absolute diameter, that the conic sections are the only curves whose diameters must always be straight lines. It is essential to an absolute diameter that the algebraic sum of its ordinates should vanish for any given absciss, the positive and negative values being equal: and the equation to any conic section may always be so transformed as that this may be effected, by making the uneven powers of y to disappear. Note to page 13, line 36.

But it is not so with lines of the 3d order; for although we may always make one of the uneven powers of y to disappear, by changing the direction of the ordinates, we cannot always make the other terms containing uneven powers of y to vanish contemporaneously. But a curve line may be always drawn which shall so divide the ordinates between it and the curve as that their algebraic sum may equal o, or that the sum of the positive products may equal the sum of the negative products of the roots.

In the case of a line of the 3d order this curvilinear diameter will be a conic section. In that of a line of the 4th order it will be a line of the 2d or 3d order, &c., as may be inferred also from the consideration that when the equation to any curve is arranged according to the powers of one of its variables commencing with the highest, the coefficient of the 2d term equated to o represents a straight line, that of the 3d term a conic section, that of the 4th term a line of the 3d order, &c.

A curve may therefore have a diameter of any order inferior to its own, according as one or more of the coefficients in its equation duly arranged is zero.

Notes to Section IV.

" *When a redundant hyperbola has no diameter, let the four roots be found of the equation*

$$a x^4 + b x^3 + c x^2 + d x + \frac{e^2}{4} = 0 \quad "$$

It is necessary to find these four roots in order to obtain the limits within which the curve exists. Now the solution of Newton's first case of equations where all the terms are present gives

$$y = \frac{e}{2 x} \pm \frac{\sqrt{a x^4 + b x^3 + c x^2 + d x + \frac{e^2}{4}}}{x}$$

which furnishes the two values of the ordinate y.

The first term of the right-hand side of this equation indicates a hyperbolic diameter, while the second term represents the ordinates bisected by this hyperbolic diameter. So that the existence and position of the curve depend upon the nature of the irrational numerator; hence the necessity of finding the four roots of the biquadratic.

When the quantity under the quadratic vinculum $= 0$, nothing but the intersection of the curve with the hyperbolic diameter is indicated in the equation which becomes

$$y = \frac{e}{2 x} \pm 0,$$

but if the irrational quantity is greater than 0, the ordinates have a real value, and corresponding points in the curve will exist; and when it is less than 0, since the irrational quantity becomes imaginary, there will be no value of y corresponding to the value of x, and consequently no curve, for, as De Lagny remarks, the imaginary term " rend toute la formule imaginaire par une espèce de contagion qu'elle communique à la partie réelle."

When we have ascertained these four roots we may proceed to determine the limits within which the curve will be found, and thence to assign the species to which any individual equation refers; for proceeds the text,

" Let the roots be AP, Aϖ, Aπ, Ap, *draw the ordinates* PT, ϖT, $\pi\mathit{l}$, *pt, then these will touch the curve, and, by touching, give the limits of the curve."* Note to page 14, line 7.

That is, since the irrational quantity already referred to, may be positive, zero, or negative, and cannot pass from the positive to the negative category, or *vice versâ*, without intermediately becoming $= 0$, we see that the point in the curvilinear diameter where the ordinate passes from real to imaginary is the limit between the existence and non-existence of the curve; so that the curve will be found on one side of this point, but not on the other. The ordinate through this point, therefore, touches the curve.

" If all the roots are real, unequal, and of the same sign, the curve consists of three hyperbolas, inscribed, circumscribed, and ambigenous, and of an oval." Note to page 14, line 11.

Such are the conditions necessary to constitute the first species. Let AP, Aπ, Aϖ, Ap, represent the four real unequal roots which, measured in the same direction from the origin A, will be of the same sign. Through P, π, ϖ, p, draw lines parallel to the axis of y. These lines will divide space into regions, and also will touch the curve. If the curve exists in the region APT, it will not exist in the region P$\pi\mathit{l}$, it will be found in the region $\pi\varpi\tau$, but will become imaginary in the region ϖpt, beyond which it will accompany the asymptotes to infinity.

Hence the portion of the curve included in the space $\pi\varpi\tau$ is necessarily an oval, for if it had infinite branches, it would be possible to draw a straight line to cut the curve in four points, in which case it would not belong to the system of lines of the third order. Since each of the asymptotes intersects the curve once, and only once, and since there is no rectilinear diameter, it necessarily follows that one hyperbola is circumscribed and one ambigenous, the third being consequently inscribed.

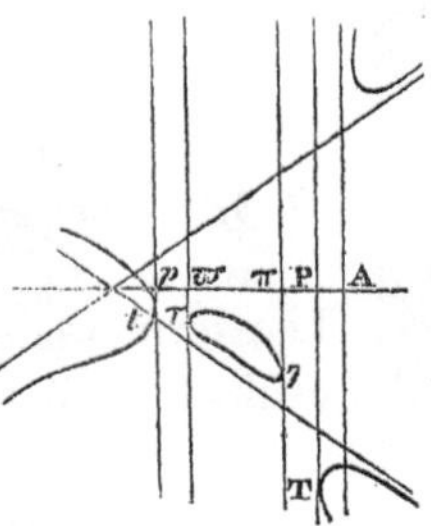

When the two greatest roots, or the two least, are equal, the oval becomes joined to the circumscribed hyperbola, and forms a node or folium, as it has been sometimes called. Nicole, who commenced writing an elaborate commentary on Newton's treatise on the enumeration of lines of the third order, which is printed in the " Mémoires de l'Académie," has fallen into the mistake of supposing that the oval is capable of uniting with the ambigenous hyperbola. It is evident from the diagram that this cannot be the case, for if it could occur, it would follow that a part of the oval might exist outside of the asymptotic triangle, which is impossible; and it is equally impossible that the ambigenous hyperbola should enter that triangle, its branch being below the asymptote, to which it is inscribed.

The circumscribed hyperbola, when it joins the oval, will always have its branches convex towards each other, otherwise a straight line might cut the curve in more than three points.

There is nothing in the remaining seven species of redundant hyperbolas sufficiently remarkable to call for particular notice.

When two roots are equal, and the other two are also equal, the curve is no longer one of the third order, but is a combination of a straight line with a conic section, for the quantity

$$\sqrt{a x^4 + b x^3 + c x^2 + d x + \frac{e^2}{4}}$$

is in this case no longer irrational, and therefore the equation

$$x y^2 + e y = a x^3 + b x^2 + c x + d$$

is divided into two equations, one representing a straight line, and the other a conic.

Note to page 15, line 26.

" Of the twelve redundant hyperbolas with one diameter."

The term ey being deficient in the equation to these curves, the ordinate will be expressed by

$$y = \pm \sqrt{\frac{a x^3 + b x^2 + c x + d}{x}}$$

Here the absciss itself is a diameter, as appears from the double sign.

Newton has enumerated twelve of these hyperbolas, and two more have been added to the number by Stirling. If in the figure an oval or conjugate point be supplied, these two additional curves will be described. The case is that in which the two lesser roots are real, while the hyperbolas consist of one inscribed and two ambigenous curves.

There are, therefore, according to the principle laid down by Sir Isaac Newton himself, fourteen redundant hyperbolas having one diameter.

In the " Ladies' Diary" for 1788 a prize question is proposed in the following terms:—" Fig. 20 in Newton's Catalogue consists of two ambigenous hyperbolas at d and δ, and one inscribed at D, without oval or conjugate point; but between these two there are five more curves, essentially different from either; two of which have been described by Mr. Stirling. It is, therefore, required to determine the other three, with an example of a numerical equation for each." In the answers which were given to this challenge, the writers have been misled into the assumption that a difference in figure constitutes a difference in species. By assuming the roots so that the vertex of the hyperbola (fig. 20) falls to the left of D, they argue that three additional curves may be added to the catalogue. From the numerical examples given, viz.,—

$$x\,y^2 = \frac{1}{4}\,x^3 + 4\,x^2 + 17\,x + 20 \ (\text{roots} - 2, - 4, - 10)$$

$$x\,y^2 = \frac{1}{4}\,x^3 + 4\,x^2 + 17\tfrac{1}{4}\,x + 22\tfrac{1}{2} \ (\text{roots} - 3, - 3, - 10)$$

$$x\,y^2 = \frac{1}{4}\,x^3 + 4\,x^2 + 18\,x + 30 \ (\text{roots} - 3, \pm \sqrt{} - 3, - 10)$$

all of which belong legitimately to Newton's species already enumerated, it is evident that these three proposed curves are not " essentially different" from the enumerated species.

" Of the two hyperbolas having three diameters."

Note to page 17, line 1.

According to what has already been shown, the hyperbolas admitting of three diameters are represented by the first case of the equations, where $e = 0$, and $b^2 = 4\,ac$.

The equation in this case, by the substitution of $\dfrac{b^2}{4a}$ for its equal (c), becomes

$$x y^2 = a x^3 + b x^2 + \frac{b^2}{4a} x + d$$

Sir I. Newton appears to have omitted to consider all the cases of the roots of this equation, and has only taken notice of the curves which result from the case in which two of the roots are imaginary. Stirling, however, has pointed out that all three roots may be real, and that then the curve admits of an oval, or of a conjugate point, according as the two lesser roots are unequal or equal. No other case can arise with reference to the roots of the equation, for it is not possible that its two greater roots should be equal, neither is it possible they should be of different signs.

<table><tr><td>Note to
page 17,
line 7.</td><td>

" Of the nine redundant hyperbolas having three asymptotes converging to a point."

</td></tr></table>

When the three asymptotes converge to a point, the triangle formed by their intersection with one another is evanescent. The coefficient, b, in the first case of the equations then becomes $= 0$, and the equation resulting is

$$x y^2 - e y = a x^3 + c x + d$$

The principal difference between these redundant hyperbolas and those which have been hitherto discussed is, that they, under no circumstances, admit of an oval, conjugate point, node, or cusp. For these can only occur within an asymptotic triangle, in the case of redundant hyperbolas. The nine species of this curve enumerated by Newton present no subject for remark.

<table><tr><td>Note to
page 17,
line 27.</td><td>

" Of the six defective hyperbolas without a diameter."

</td></tr></table>

In the first case of the equations, when the term $a x^3$ is supposed to be negative, the equation solved for y, gives

$$y = \frac{e}{2x} \pm \frac{\sqrt{-a x^4 + b x^3 + c x^2 + d x + \dfrac{e^2}{4}}}{x}$$

From which it appears that the curve admits of only two infinite branches, and it is, therefore, called defective by Newton; for the quantity under the quadratic vinculum becomes impossible as soon as the first term exceeds the remainder of the expression in the numerator, and cannot again become possible for any value of x.

Since the curve which has no rectilinear diameter cuts its asymptote in one point, the infinite branches of these six defective hyperbolas lie on different sides of their asymptote. It is evident, also, that they admit of a hyperbolic diameter bisecting the ordinates terminated by the curve, on which there may occur an oval, node, &c., according as the roots of the irrational portion of the equation are equal or unequal to one another.

The case where two of the roots are impossible, and at the same time b and $d = $ o, is remarkable, for then the curve has a centre, though it has no diameter. The same curve when its equation by a transformation of axis assumes the form

$$y^3 = a x^2 - x^3$$

has been called by some writers a circle of the second order, because its equation may be conceived to be derived from a property analogous to that of the circle. Thus, as the equation to the circle whose diameter is (a), is derived from the proportion,

$$x : y :: y : a-x,$$

so in circles of the higher orders the equations are derived from the proportion,

$$x^n : y^n :: y : a-x.$$

If $n = 2$, we have,

$$x^2 : y^2 :: y : a-x, \text{ or } y^3 = a x^2 - x^3,$$

an equation to the defective hyperbola under discussion. If $n = 3$, $y^4 = a x^3 - x^4$, an equation to the lemniscate, and generally if (n) is an even number the curve will be infinite, if n be odd, the curve will be closed.

Note to
page 18,
line 27.

" Of the seven defective hyperbolas with a diameter."

These curves differ from the last six species only by the deficiency in the equation defining them of the term $e\,y$. Since they possess an absolute diameter, their infinite branches will not intersect the asymptote, and will therefore lie on the same side of it. The cissoid of the ancients, which is one of these defective hyperbolas, is a curve of historical interest, having been invented either by Diocles or Hippias, for the purpose of finding two continued mean proportionals between two given lines.

In the *Arithmetica Universalis*, Newton has given a method of describing this curve by the continued motion of a sliding ruler. The curve may also be described by the vertex of a conic parabola made to roll upon an equal fixed parabola, their vertices being supposed coincident at the commencement of the motion. It is worthy of remark that any other point than the vertex in the rolling parabola, will trace out one of Newton's defective hyperbolas. (*See* fig. 45), having a double point in the like point of the fixed parabola.

The cissoid may also be described in the following manner by points.

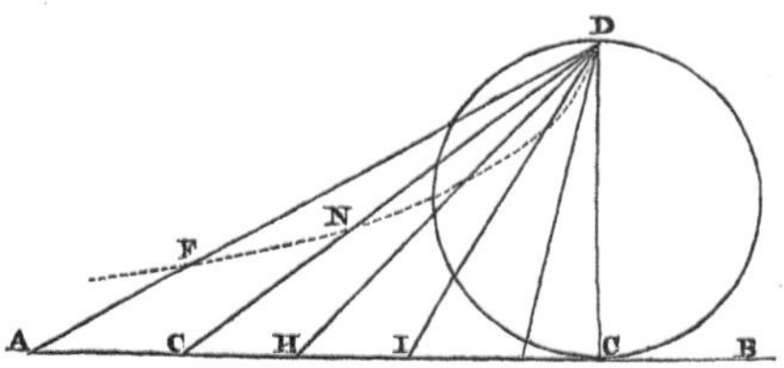

The infinite line A B being drawn perpendicular to the diameter C D, and touching the circle in C, let any number of lines be drawn from D to cut A B. Then measuring from the points of intersection with A B, lines equal to the corresponding chords, cut off the circle, as A F, G N, &c., the extremities of these lines will lie in the curve of a cissoid.

It is evident that all cissoids referred to the same axis are similar figures.

Note to
page 19,
line 23.

" Of the seven parabolic hyperbolas having no diameter."

The equation to these curves differs only from that already discussed in the case of the redundant hyperbolas, inasmuch as the term $a\,x^3$ is deficient. It then becomes,

$$xy^2 - ey = bx^2 + cx + d \quad . \quad . \quad . \quad (1)$$

whence,

$$y = \frac{\dfrac{e}{2x} \pm \sqrt{bx^3 + cx^2 + dx + \dfrac{e^2}{4}}}{x} \quad . \quad . \quad . \quad (2)$$

We draw the same inference from the expression for y when x is very small, as in the curves referred to, viz., that the axis of y is an asymptote having two hyperbolic branches, one on each side of it, in contrary directions, and that this asymptote is intersected by one of the branches at a point above or below the origin, according as (e) is positive or negative, at a distance $= \dfrac{d}{e}$

But if we expand y in a series for large values of x, we shall no longer find rectilinear asymptotes, although the possibility of values for y continues to infinity. For it is evident, since the irrational quantity in the second member of the equation, when x approaches infinity becomes $\pm\sqrt{bx}$, that the branches of the curve admit only of a curvilinear asymptote, and are consequently parabolic.

If in equation (2), x is supposed to be negative, the ordinate becomes imaginary as soon as the irrational portion of the equation becomes negative, and hence there are no more infinite branches than the four already spoken of, namely, two hyperbolic having a rectilinear asymptote, and two parabolic, having a conic parabola for their asymptote whose parameter is b, and whose diameters are parallel to the axis of x.

Expanding y in a series we have,

$$y = \pm\sqrt{bx} \pm \frac{c}{2\sqrt{bx}}\&c.,$$

which shows that the asymptote parabola will be inside or outside of the parabolic branch, according as c is positive or negative.

That this asymptote can only cut the curve twice, may be thus shown. At the point where the curve and asymptote intersect, $y = \pm\sqrt{bx}$, whence $\dfrac{y^2}{b} = x$, is the corresponding value of the abscissa, substituting which value in the equation (1) we obtain another equation,

$$y = \frac{-be \pm \sqrt{b^2e^2 - 4bcd}}{2c},$$

which does not give more than two values for y, and therefore there will be no more than two points of intersection. When $4\,c\,d > b\,e^2$, the equation determining the points of intersection becomes impossible, that is, the parabola will not intersect the curve, but will lie entirely within its branches. When the parabola intersects the curve twice, the points of intersection will lie on the same or on opposite sides of the absciss, according as c is positive or negative in sign.

If the equation (1) be solved for x, it shows a parabolic diameter, bisecting chords parallel to the axis of x.

Note to
page 20,
line 28.

" Of the four parabolic hyperbolas having a diameter."

The hyperbolo-parabolic curves which have a rectilinear diameter, as enumerated by Newton, are four in number. Their species being furnished by the cases in which the roots of the equation,

$$b\,x^2 + c\,x + d = 0,$$

are (1) impossible, (2) equal and of the same sign, (3) unequal and of the same sign, (4) of different signs.

It is remarkable that neither Newton nor his commentator Stirling should have noticed that when the roots of this equation are possible, and (whether equal or unequal) of the same sign, the curve will differ according as that sign is positive or negative. So that there are in reality six hyperbolo-parabolic curves with a diameter.

Thus when the roots are positive the equation will be of the form,

$$y = \pm \sqrt{b\,x - c + \frac{d}{x}},$$

an equation giving possible values for y, only on the assumption that x is measured on the positive side of the origin, for when x is taken negatively, the values of $\pm\,y$ are all imaginary.

But if the roots are negative,

$$y = \pm \sqrt{b\,x + c + \frac{d}{x}},$$

and when x is negative this becomes,

$$y = \pm \sqrt{-bx + c - \frac{d}{x}},$$

indicating certain limits between which a closed curve will exist on the negative side of the origin. The figure will be as shown.

When the negative roots are equal, the oval becomes a conjugate point.

This curve is the subject of a paper in the Transactions of the Royal Society in the year 1736, by Mr. Edmund Stone, a learned mathematician of the time, who seems not to 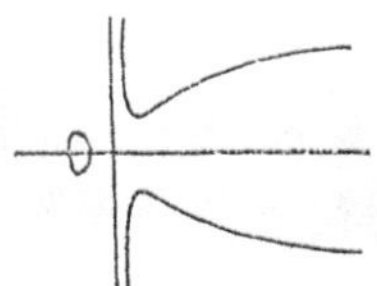 have been aware of the previous notice of the same curve by others. It appears to have been first observed upon and figured by Nicole in an article on Newton's Theory of Shadows, published in the Mémoires de l'Académie for the year 1731. The same curve was also known to N. Bernouilli, and in his " Usage de l'Analyse de Descartes," the Abbé de Gua remarks the omission of it by Newton in his enumeration.

Of the hyperbolisms of the conic sections.

Note to
p. 21,
line 16.

When the first case of the equations becomes still further simplified by the deficiency of the terms $a x^3$ and $b x^2$, and is of the form,

$$x y^2 + e y = c x + d \quad \cdot \quad \cdot \quad \cdot \quad (1)$$

Newton calls the curves which are defined by it, hyperbolisms of the conic sections, because they may be generated from the ordinate of the hyperbola ellipse or parabola; thus if A B represent a given line, the locus of the extremity of the ordinate tracing out the hyperbolism is found by multiplying the ordinate M P by the given line A B, and dividing the product by the common absciss A M.

" *In this manner a straight line is turned into a conic hyperbola,*"

Note to
p. 21,
line 22.

thus, if the straight line be represented by $y = x + c$, and this ordinate be multiplied by the given line a, and divided by the common absciss x, the equation to the new curve is $y = \dfrac{a\,(x + c)}{x}$ which represents a hyperbola.

In like manner where the ordinate of a conic section is,

$$\frac{-e \pm \sqrt{e^2 + 4\,dx + 4\,c\,x^2}}{2\,m},$$

this being multiplied by a given straight line m, and divided by the common absciss x, becomes

$$\frac{-e \pm \sqrt{e^2 + 4\,dx + 4\,c\,x^2}}{2\,x},$$

for the ordinate to the hyperbolism.

Case 1. Where $e > 0$, and c is positive.

The axis of y is an asymptote having an infinite branch lying on each side of it, extending in contrary directions, as appears from expressing y in a series for small values of x. If y be expanded in a series converging for large values of x, we have,

$$y = \pm \sqrt{c} \pm \frac{d \mp e\sqrt{c}}{2\,x\,\sqrt{c}} + \&\text{c.} \quad . \quad . \quad . \quad (2)$$

which indicates two parallel asymptotes above and below the axis of x, at distances $+\sqrt{c}$, and $-\sqrt{c}$. Whence it is evident that when c is negative these parallel asymptotes become imaginary, and when $c = 0$, that they coalesce and form one line.

The equation (1) solved for y giving,

$$y = -\frac{e}{2\,x} \pm \sqrt{c + \frac{d}{x} + \frac{e^2}{4\,x^2}} \quad . \quad . \quad . \quad (3)$$

shows a hyperbolic diameter having infinite branches of the curve on each side of it, whatever be the sign of x, also when $x = 0$, the ordinate is an asymptote having a branch on each side of it. There are therefore six branches, and the curve is consequently a hyperbolism of the hyperbola. Since the parallel asymptotes are also parallel to the axis of x, they can never be intersected by the curve.

Case 2. Where $e = 0$, and c is positive.

Here the ordinate $= \pm \sqrt{c + \dfrac{d}{x}}.$

The curve, since it has an absolute diameter, will not intersect the asymptote δd (fig. 68); and the infinite branches will always lie on the outside of the parallel asymptotes when x is positive, but on the inside of them when x is negative. This is evident, because the equation to the asymptotes is $y = \sqrt{c}$, and $\sqrt{c}$ is always less than $\sqrt{c + \dfrac{d}{x}}$, greater than $\sqrt{c - \dfrac{d}{x}}.$

Case 3. Where $e > 0$, and c is negative.

The ordinate becomes $= - \dfrac{e}{2x} \pm \sqrt{-c + \dfrac{d}{x} + \dfrac{e^2}{4x^2}}.$

There is here only one pair of infinite branches cutting their asymptote; the parallel asymptotes become imaginary; the curve is a hyperbolism of the ellipse, having hyperbolic diameters.

Case 4. Where $e = 0$, and c is negative.

The infinite branches lie on the same side of the asymptotes, for there is now an absolute diameter. The form of the curve is capable of an infinite variety of figure, from a straight line coinciding in the asymptote to an elongated cusp-like curve, which occurs when d greatly exceeds c.

If the foci of the ellipse from which this curve is conceived to be generated coincide, so that the ellipse becomes a circle, the resulting hyperbolism is the same as the curve called the Witch, the invention of which has been generally attributed by modern writers to the celebrated female professor of mathematics in Bologna, Donna Maria Agnesi, A.D. 1740. It, however, does not appear, in the *Istituzioni Analytiche* of the authoress, that she advances any claim to the invention; nor does she appear to have been responsible for the unmeaning name attached to the curve, for she speaks of it as the curve " volgarmente detto la Versiera." Her work, which abounds with examples of interesting and curious problems, was translated into English, and published, after the death of the translator, Mr. Colson, by Baron Maseres, in 1801.

According to Agnesi (*Inst. Anal.* Book I. Sect. 4), the Witch is thus generated:—"The semicircle A D C being given, required a point outside it, such that, drawing a perpendicular M B to the diameter, cutting the circle in D, we may have $AB : BD = AC : BM$. And because there will be an infinite number of points that will satisfy the problem, the locus of these points is required."

Thus, if $AC = a$, $AB = x$, $BM = y$, $BD \sqrt{ax - x^2}$,

$$x : \sqrt{ax - x^2} :: a : y,$$ therefore

$$y = a \sqrt{\frac{a - x}{x}};$$

which equation is the same as Newton's, where $d = a^3$ and $c = a^2$.

Case 5. When $e > 0$, and $c = 0$.

The equation is now

$$xy^2 + ey = d; \text{ whence } y = - \frac{e}{2x} \pm \sqrt{\frac{e^2}{4x^2} + \frac{d}{x}}$$

When x is positive, it is evident there are two infinite branches on each side of the axis of x, extending in the same direction, and at an infinite distance coinciding with it. This apparent anomaly is explained by the consideration that since $c = 0$, the two parallel asymptotes defined by $y = \pm \sqrt{c}$, have come together, and the axis of x is consequently a double asymptote.

The axis of y is also an asymptote, having infinite branches on different sides of it; the curve, which is a hyperbolism of the parabola, is the only one of the third genus which has four branches only, all of the hyperbolic sort.

Case 6. When $e = 0$, and $c = 0$.

The equation is $xy^2 = d$.

Unlike the hyperbolism of the ellipse, this curve is susceptible of but one figure when referred to axes of given inclination. The simplicity of its equation renders it useful in the construction of equations by means of the intersection of curve-lines. The hyperbolism of a parabola with a diameter is called by some writers the cubic hyperbola.

The Second Case of the Equations.

Note to p. 23, line 24.

The curve indicated by the equation

$$xy = ax^3 + bx^2 + cx + d$$

furnishes no subdivisions into species, and is easily discussed; for, since it gives

$$y = ax^2 + bx + c + \frac{d}{x},$$

it is plain that for any value whatever, positive or negative, of the absciss, only one value of the ordinate can exist, on the supposition which is assumed, viz., that the ordinate is parallel to a rectilinear asymptote.

When $x = 0$, y is infinite, and therefore the axis of y is an asymptote, having a branch extending in one direction; and when $- x = 0$, there will be another branch in a contrary direction, but these branches will not unite.

When $x = \pm \infty$, $y = ax^2$, or the two remaining branches of the curve approach to a conic parabola, and as the values of x are continuous from ± 0 to $\pm$ infinity, the parabolic branches unite with the hyperbolic.

The ordinate to the parabolic asymptote is

$$ax^2 \pm bx + c,$$

that to the curve is

$$ax^2 \pm bx + c \pm \frac{d}{x},$$

the upper sign being taken when the roots are negative; whence it appears that the asymptote on the positive side lies above the curve, and on the negative side below the curve, its position being governed by the sign of $\dfrac{d}{x}$, the first term containing x in the denominator. It can therefore never cut the curve.

Newton calls this single instance of a curve, corresponding to his second case of equations, the " Cartesian parabola," because, as he says, Des Cartes by its assistance constructed an equation of six dimensions.

The trident, as it is also called, may be readily described by observing that the ordinate consists of the sum or difference of the ordinates of the parabola

$$y = a\,x^2 \pm b\,x + c,\text{ and the hyperbola } y = \pm\,\frac{d}{x}.$$

Des Cartes gives a method of describing the trident by continuous motion. (See also Maclaurin's *Geometria Organica*.)

———

Note io
p. 24,
line 1.

Third Case of the Equations.

The five species enumerated as referable to the equation

$$y^2 = a x^3 + b x^2 + c x + d$$

differ only in the finite part of the curve; for, since

$$y = \pm \sqrt{a x^3 + b x^2 + c x + d}$$

when x is very great, the terms following $a x^3$ may be neglected; and it follows that the semicubic parabola represented by $y = \pm \sqrt{a x^3}$ is the asymptote to all the divergent parabolas.

Newton calls this curve the " Neilian parabola," in compliment to Mr. W. Neil,* who first showed how to rectify the curve by means of the quadrature of a parabolic space.

---

* Neil was born in 1637, was educated at Wadham College, Oxford, and died in 1670. His discovery of the method for rectifying the semicubic parabola was made when he was only nineteen. (See *Phil. Trans.* for 1673.) Neil's priority in this discovery was disputed, but in Newton's opinion without reason.

"*If two roots are impossible, a pure bell-shaped parabola will result (figs. 78, 79).*"

It may, however, happen that when two roots are impossible, the figure of the curve in question is not bell-shaped (campaniformis).

Thus, suppose the equation made up of the factors

$$x^2 + h^2 = 0, \; x + k = 0,$$

so that it is

$$y^2 = x^3 + kx^2 + h^2 x + kh^2,$$

differentiating and assuming $dy = 0$, we have

$$3x^2 + 2kx + h^2 = 0, \text{ or } x^2 + \frac{2}{3}kx = -\frac{h^2}{3},$$

solving the quadratic,

$$x = -\frac{1}{3}k \pm \sqrt{\frac{1}{9}k^2 - \frac{1}{3}h^2},$$

which, when real, shows that there is a pair of maximum and a pair of minimum values for y, and therefore the curve cannot be bell-shaped. The values obtained for x are those of the abscisses belonging to the points at which the tangent to the curve is parallel to the axis. Whenever, therefore, the irrational quantity is real, the parabola will not be bell-shaped. If $\frac{1}{9}k^2 = \frac{1}{3}h^2$, there will be a point of inflexion when $x = -\frac{1}{3}k$, and the tangent at this point will be parallel to the axis.

If a solid generated by the revolution of a semicubic parabola about its axis be cut by a plane, each of these five parabolas will result from the sections; for when the cutting plane is oblique to the axis, but falls below the vertex, the section is a diverging parabola with an oval. When the plane passes through the vertex, this oval becomes a point; when the plane both cuts and touches the surface of the solid, the section is a nodate parabola; when the plane falls above the vertex, either parallel or oblique to the axis, the section is a pure parabola; and when the plane passes through the axis, the section is the semicubic.

Note to
p. 24,
line 25.

Fourth Case of the Equations.

$$y = ax^3 + bx^2 + cx + d.$$

Here, again, as in the case of the trident, we can have no more than one value for y, whatever may be the value of x, positive or negative. This case evidently admits of no subdivision into species, for when the equation

$$y = ax^3 \pm bx^2 + cx \pm d = 0$$

has real and unequal, or real and equal roots, &c., nothing is indicated except that the curve cuts or touches the axis of x in certain points. The branches are evidently parabolic, and extend on each side of the axes; the curve is called the cubic parabola. As the semicubic was the first curve which was rectified, so the cubic parabola will probably be the last, a solution of the problem of its rectification being still a desideratum.

Note to
p. 24, last
line.

" Thus altogether the species are 72."

De Gua has remarked, in his review of Newton's enumeration, that before commencing it some definition should have been given to fix the meaning intended to be attached to the word " species;" and it would seem that the want of some such definition has been the cause of much criticism on the part of subsequent writers which might have been avoided. It would, perhaps, have been better to have classified the curves into species and subdivisions of species, in which case it might be easily shown that the species are at most not more than six or seven; provided that Newton's own principle of classification be adopted, namely, the relation to one another of the roots of the equations considered, as referred to axes one of which is parallel to an asymptote. De Gua further observes that Newton has proceeded solely on the principle of considering the number, position, and equations respectively of the maxima and minima of x without adopting the same course relatively to the maxima and minima of y, although the axis of x and that of y are equally determinate straight lines, and in a singular position. And that if he had made the same observations on the maximum and minimum values of y as he has done upon those of x, the species 1, 2, 3, 6, 10, 14, 15, 16, 22, and 28, might have been further subdivided, and consequently a much larger number of species enumerated. It is also a common observation of modern writers on algebraic geometry and the differential calculus, that Newton's enumeration is far from complete, and might be extended so as to include many more species. To which it may be replied, that his principle of classification is entirely arbitrary, and that if we adopt that principle, it is in the highest degree improbable that the number of species could be increased, seeing that the ingenuity and research of mathematical writers for more than a century have failed to add more than six additional curves to the total number. It is worthy too of remark, that these six curves are different from already enumerated species, only by the addition of an oval or conjugate point.

Both Cramer, and Euler in his "Analysis Infinitorum," have given a classification of lines of the third order, depending upon a principle entirely different from Newton's, namely, upon a consideration of the nature and direction of the infinite parts of the curves. According to Cramer, there are fourteen different species, and, according to Euler, sixteen different species. Neither of them considers the existence of a conjugate oval or point as sufficient to constitute a separate species. This view of the subject may be less liable to logical objection than that which Newton has adopted, but it is not so well calculated to give a general idea of the varieties of which the different curves are susceptible.

In more modern times the same subject has been taken up with much elaboration of research by mathematicians in Germany, especially Plücker, Mobius, Peters, Krause, Beer, &c., but it is impossible to institute a comparison between the classification of curves adopted by these writers and the enumeration of Newton, because the former proceeds upon the assumption of a different system of co-ordinates, and a different notion of the genesis of curve lines. Plücker gives a catalogue of 219 curves of the third order. According to the German writers, a curve is to be considered as the envelope of a moveable right line, and is classed according to the number of tangents to it which can be drawn through any given point. A line of the third order is thus a curve of the sixth class, because six tangents may be drawn to it from one point.

Dr. Waring undertook the herculean task of the enumeration of lines of the fourth order upon the Newtonian system, and the result of his labours is published in his " Miscellanea Analytica," printed at Cambridge, 1762. He reduces all lines of the fourth order to twelve cases of equations, beginning with the biquadratic parabola, of which there is only one species, and ending with the curve possessing four asymptotes, of which no less than 72,480 varieties are specified. The total number of species he calculates at 84,551. An interesting work on curves of the fourth order, con-taining many plates, entitled, " Tabulæ curvarum quarti ordinis symmetricarum asymptotis rectis et lineâ fundamentali rectâ præditarum," was published in 1852 by Professor Augustus Beer, at Bonn, which is more likely to be useful to the student than Dr. Waring's catalogue.

Notes to Section V.

Note to
p. 25,
line 2.

The generation of curves by shadows.

The remarkable theory so briefly enunciated in this fifth section appears to be substantially the same as that which is discussed at greater length in the 22d lemma of the Principia, in which it is proposed to ' transmute' any rectilinear or curvi-linear figure into another of the same analytical order by means of the method of projections. This theory, as Newton has re-marked, is of considerable value in the solution of problems* which scarcely admit of being otherwise solved, and he has him-self made use of it in Prop. 25, 26, Book I. of the Principia.

We premise the discussion of the principles on which the genesis of curves by shadows is founded, by the following definitions.

---

* Inservit autem hoc Lemma solutioni difficiliorum Problematum trans-mutando figuras propositas in simpliciores. PRIN.

1. A plane of infinite extent being supposed, on which is delineated the curve which is proposed to be projected, this plane is called the " plane of the base."

2. A second plane parallel to the plane of the base, is called the " horizontal plane."

3. A third plane cutting both these planes at any given angle, is called the " plane of projection," from its receiving the projection of the curve delineated on the first-named plane.

4. A fourth plane is to be supposed, parallel to the third plane, at a finite distance from it.

5. The line of intersection between the horizontal plane and the plane of projection is called the " horizontal line."

6. The line of intersection between the plane of the base and the fourth plane is called the " directrix." Its position with regard to the curve to be projected determines the species of the projection.

7. The line of intersection between the plane of projection and the plane of the base, is called the " line of the base."

Cor. The horizontal line is parallel to the line of the base.

8. If an infinite straight line is supposed to revolve about a fixed point P, situated in the line of intersection between the horizontal plane and the fourth plane, and is guided along the course of the curve delineated on the plane of the base, it will, by its intersection with the plane of projection, describe the projection of the given curve.

9. A plane supposed to be drawn through this pole P, and perpendicular to the before-mentioned planes is called the " vertical plane," and its intersection with the horizontal plane is the " axis of projection;" the point in which the axis of projection meets the horizontal line is called the " centre of projection."

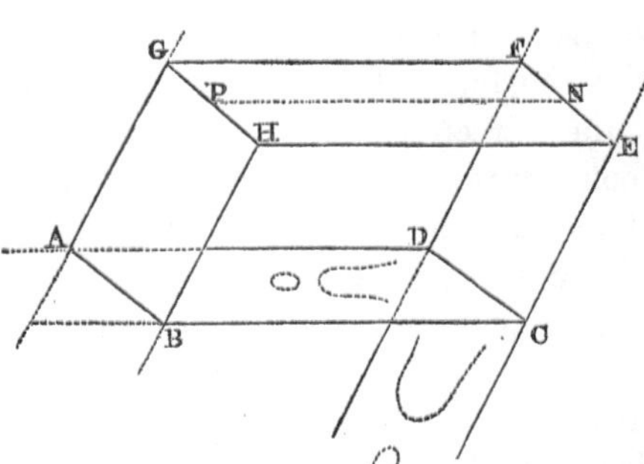

Thus in the annexed figure A B C D represents the plane of the base, G H E F the horizontal plane, D C E F the plane of projection, A B H G the fourth plane, A B the directrix, C D the line of the base, E F the horizontal line, or, as it is sometimes called, the anti-directrix, P the pole, PN the axis of projection, and N the centre of projection.

It is evident that several points in a straight line can only be projected on a plane by as many points also in a straight line, and thence it follows that the outline of the shadow of any curve may be cut by a straight line precisely in as many points as the curve, of which it is the shadow, may be cut by a straight line; and therefore the proposed curve and the projected curve are of the same degree. Moreover, the projection of the tangent or the asymptote, to any point whatever, is always either a tangent or an asymptote to the projection of this point; and the projection of the axis of any curve will be the axis of the new curve.

The curve may lie on either side of the directrix, or partly on one side, partly on the other. When this latter case arises, or when it touches the directrix, the point or points projected which cut or touch the directrix, go off to infinity. Thus, for example, suppose a circle delineated on the plane of the base, it is obvious that while it is placed to the right of the directrix, its projection will be thrown on the plane of projection below the line of the base DC. When placed to the left of AB, its projection will be thrown on the plane above FE, in either case being an ellipse. If, however, it touches the directrix, the projected curve will have infinite branches ultimately becoming parallel, and the curve will be a parabola; and if it is cut by the directrix, there will result four infinite hyperbolic branches, of which two will be above and two below the line of the base.

Suppose now the curve delineated on the plane of the base is a semi-cubic parabola, lying wholly between the directrix and the line of the base, but without touching either, and having its axis at right angles to these lines, its ordinates parallel to them. The projection will take the form of a cissoid. If the cusp of the semi-cubic touch the directrix, the projected curve will consist of four hyperbolic branches, converging to one double and one single asymptote, viz., the cubic hyperbola; and if the curve cuts the directrix in two points, each point of intersection will be projected as an asymptote, while the two branches of the semi-cubic itself will be projected towards a third asymptote, and the projection will constitute a redundant hyperbola, two of the hyperbolic curves being inscribed, and the third circumscribed and cusped (fig. 24).

We have here supposed the simplest case which can occur, namely, that in which the axis is normal to the ordinates and to the directrix, the curve itself having no points of inflexion or conjugate oval. The case in which the curve is more complicated and the directrix oblique to the axis, will be hereafter adverted to.

We now proceed to show how the equation to the projection may be obtained from that of the generating curve, or *vice versâ*. Assuming rectangular co-ordinates, let it be required to compare the curves (either of which may be considered as the projection of the other,) ATB, AEB, where S represents the pole, LSCI,

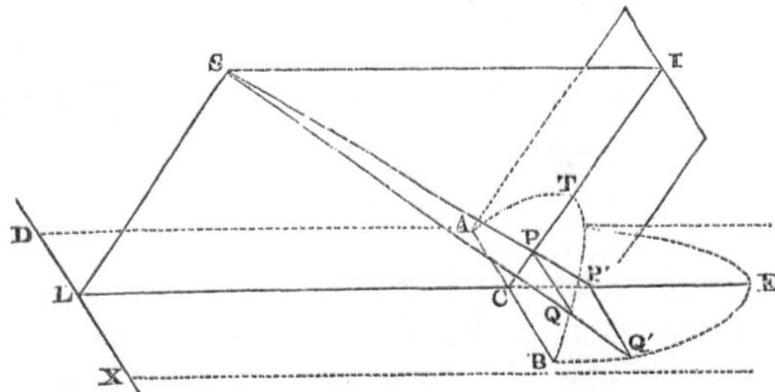

the vertical plane, Q, Q′, the points in the two curves to which the ordinates PQ, P′Q′, belong. Let I be taken as the origin in one curve, and L in the other. Let the axis CE of one curve be the projection of the axis CT of the other. Then by sim. tri. SPI, CPP′, and the equality SI = LC, CI = SL,

$$(LC + CP') LP' : SI :: SL : IP,$$

and calling the known lines SI and SL, m and n,

$$x : \pm m :: \pm n : x'.$$

Again, by sim. tri. SPQ, SP′Q′, and the equality SI = LC,

$$SP' : SP :: (LP' : SI) :: P'Q' : PQ,$$

or $$x : \pm m :: y : y',$$

whence we have the two values $x' = \dfrac{mn}{x}$, $y' = \dfrac{my}{x}$, which, substituted in the equation of the given curve, will produce that of its projection. For example, let the equation given be

$$y'^2 = a\,x'^3 + b\,x'^2 + c\,x' + d,$$

the equation to the projection will be

$$y^2 x = \frac{d}{m^2}\,x^3 + \frac{c\,n}{m}\,x^2 + b\,n^2 x + a\,n^3 m.$$

This form of the equation shows that the projection will be some curve of the third degree, not one of the divergent parabolas.

The proposition we are now to discuss is, that as the projection of the circle generates all conic sections, so that of the five divergent parabolas will generate every one of the lines of the third order.

Assuming that the curve whose projection is sought is delineated upon the plane of the base, the inclination of which, with the plane of projection, is given, and that its position, with reference to the directrix, is fixed in each case; then it will be seen that it is sufficient to know this position, in order to determine the species of the projected curve. It is evident that all the points of the curve will be projected on the plane of projection, save and except those points which are in the directrix itself, which latter will not be projected at all. The points infinitely near this line will be projected to an infinite distance, that on one side, in the direction of the positive x's, that on the other, in the direction of the negative x's. Therefore, when the directrix cuts the curve, the projections of the points immediately adjoining this line take the form of infinite hyperbolic branches to the right and left of a common asymptote, which is the projection of the tangent at the point intersected. But if the directrix only touches the curve, the infinite branches will be both positive or both negative in direction, and (except in the case where the point touched is a cusp) parabolic.

In the application of the foregoing remarks to the case of the projections of the five divergent parabolas, we shall first consider all the projections which can be furnished by the divergent parabola with an oval. (Fig. 75.)

Case 1. Where D X, the directrix, is parallel to the ordinates of the curve.

1. Let D X cut the axis outside the curve; the projection will take the form of hyperbolic branches on the same side of their asymptote, and with an oval at their convexity. (Species 39, fig. 48.)

2. Let D X touch the oval; the last projection will remain, with the exception that the oval will be converted into a parabola. (Species 55, fig. 63.)

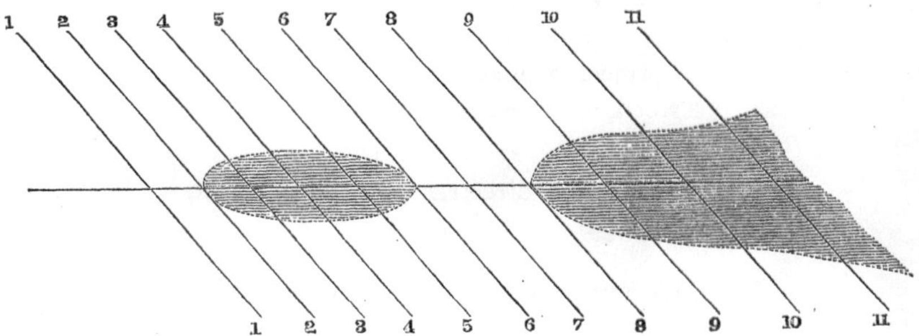

3. Let D X cut the axis in unequal parts, as at line 3. The conchoidal hyperbola remaining as before, the oval will be projected as two hyperbolas in the opposite angles of the two asymptotes. (Species 21, fig. 32.)

4. Let D X cut the oval in half; if tangents be drawn to the points where D X intersects, they will be parallel to each other; therefore their projection will be that of two asymptotes intersecting each other in the third asymptote; three hyperbolas. (Species 31, fig. 41.)

5. Let D X cut the axis at 5; the asymptotes will form a triangle. (Species 20, fig. 31.)

6. Let D X touch the oval. (Species 56, fig. 65.)

7. Let D X cut the axis between the oval and vertex of the curve. (Species 40, fig. 49.)

8. Let D X touch the parabola; the projection takes the form of a curve not enumerated by Newton or by Stirling, having an oval on one side of the asymptote and a hyperbolo-parabolic curve on the other.

9. Let D X cut the parabola, but in the part where the curve is concave to the axis; the projection gives a species which has been also overlooked by Newton, though supplied by Stirling— namely, a curve comprising three hyperbolas, with an oval in the asymptotic triangle, two of the hyperbolas ambigenous and one inscribed. To obtain the projections of the asymptotes, draw a pair of tangents from the points where D X intersects the curve; these will intersect in the axis, and also each cut the curve once.

10. Let D X cut the parabola through the points of inflexion; the projection will also be a curve omitted by Newton, viz., three inscribed hyperbolas with an oval.

11. Let D X cut the parabola where the curve is convex to the axis; the projected curve will consist of one circumscribed and two inscribed hyperbolas. (Species 10, fig. 22.)

Case 2. When the directrix is oblique to the ordinates of the delineated curve.

1. Let D X cut the parabola (without cutting or touching the oval) in any point except one of inflexion; a tangent to that point will be projected as an asymptote to the projected curve, which will be a defective hyperbola without a diameter, having hyperbolic branches on each side of this asymptote, the point where the branches intersect their asymptote being one of inflexion. (Species 33, fig. 43.)

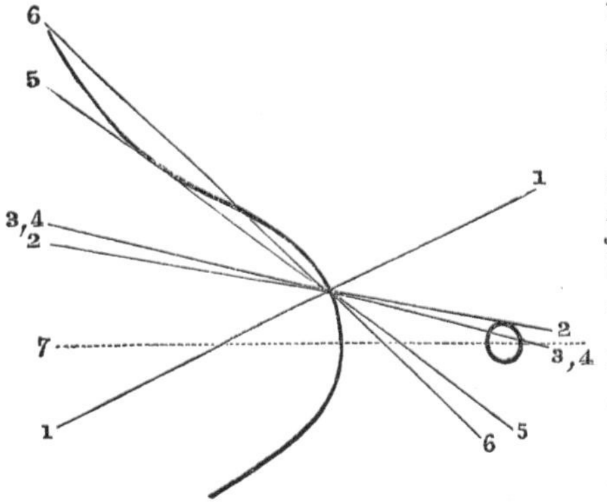

2. Let D X touch the oval, the other conditions remaining as in the last case; the serpentine hyperbola remains, the oval becomes a parabola. (Species 52, fig. 60.)

3. Let D X cut the oval and the parabola also, and conceive tangents drawn from the points of intersection with the oval, so as to meet the tangent to the point in the parabola; the tangents

will be projected as asymptotes, and the projection will consist of two inscribed hyperbolas and one serpentine hyperbola. (Species 9, fig. 20.)

4. If the two tangents to the oval meet each other in the same point as that in which each intersects the third tangent, this point of intersection will be projected as a point of intersection of the three asymptotes; and the species in this case is the 26th, fig. 36.

5. Let DX, still cutting the parabola, also touch it; the projection will consist of both hyperbolic and parabolic branches and an oval, and the curve will have no diameter; the species will be the 46th, fig. 54.

6. Let DX cut the parabola in three points; tangents drawn through these points will be projected as three asymptotes, making a triangle within which will be the projected oval; the resulting species will be the first (fig. 7), because it will consist of one circumscribed, one inscribed, and one ambigenous hyperbola; the infinite branches of the parabola being projected in the form of a circumscribed hyperbola, the part of the parabola which cuts neither of the tangents being projected as an inscribed hyperbola, and the remaining part being projected as an ambigenous hyperbola.

7. If DX coincides with the axis of the curve, the asymptotes of the projected curve intersect in a point, and the species is the 27th, fig. 37.

We have next to consider the projections of the punctate parabola.

Case 1. Where the ordinates of the delineated curve are parallel to the directrix.

1. Let DX cut the axis outside P, the conjugate point; the species will be the 43d, fig. 52.

2. Let DX cut the axis through the conjugate point; the species will be the 63d, fig. 71.

3. Let DX cut the axis in a point between P and the vertex; the species will be the 44th, fig. 53.

4. Let it touch the parabola; species not enumerated by

Newton or Stirling: a hyperbolo-parabolic curve with conjugate point.

5. Let it cut the parabola; species 13, fig. 25, and 25, fig. 35.

Case 2. Where DX is not parallel to the ordinates of the given curve.

Let P be the conjugate point, T a point in the parabola on which DX is conceived to revolve as on a pole.

1. If DX passes through P, the conjugate point, P is not projected at all; the equation consequently loses two dimensions, and the species is a hyperbolism of the ellipse. (Species 61, fig. 69.)

2. If DX passes on either side of P without cutting the curve, the projection will be a serpentine hyperbola. (Species 36, fig. 47.)

3. If DX touch the parabola, species 49, fig. 57.

4. If DX cuts the parabola in three points, species 4, fig. 12.

5. If the axis coincides with DX, the species is 62, fig. 70.

Projections generated by the pure Divergent Parabola.

Case 1. Where the ordinates of the given curve are parallel to DX.

1. Let DX cut the axis outside the curve; the projection will be species 45, figs. 52, 53.

2. Let DX touch the parabola in the vertex; the projection will be species 53, fig. 61.

3. Let DX cut the parabola in two points where the curve is concave to the axis, but so that tangents to the points of intersection meet in the axis, and also cut the infinite branches; the projection will consist of two ambigenous and one inscribed hyperbola. (Species 15, fig. 26.)

4. Let DX cut the curve through the points of inflexion; the projection will consist of three inscribed hyperbolas, corresponding to species 22, 23, or 32, figs. 33, 34, 42.

5. Let DX cut the curve beyond the points of inflexion; the projection will be species 14, fig. 25.

Case 2. When the directrix is not parallel to the ordinates.

1. Let DX cut the curve only once; the projection will be species 37, fig. 46.

2. Let DX touch and cut the curve; the projection will be species 50, figs. 57, 58.

3. Let DX cut the curve in three points. (Species 5, figs. 12, 13.)

4. Let DX coincide with the axis. (Species 38, fig. 47.)

Projections generated by the Nodate Parabola.

Case 1. When the directrix is parallel to the ordinates of the given curve.

1. Let DX cut the axis outside the curve. Projection, Species 41, fig. 50.

2. Let it touch the folium. Projection, Species 54, fig. 62.

3, 4, 5. Let it cut the folium, the projection will be three inscribed hyperbolas, two of which intersect one another. Species 18, 19, 30, figs. 29, 30, 40.

6. Let DX cut the point of intersection of the given curve. The projection will be a hyperbolism of the hyperbola. Species 60, fig. 68; for the tangents to the double point will be projected as two parallel asymptotes.

7. Let DX cut the curve inside the node. Species 11, fig. 23.

Case 2. Where the directrix is oblique to the ordinates.

Suppose DX to revolve about T, a point in the curve, as a pole.

1. Let it cut the curve once. Species 34, fig. 45.

2. Let it touch the folium. Species 51, fig. 59.

3. Let it cut the folium. Species 8, fig. 17, or Species 7, fig. 16.

4. Let it cut the double point. The tangents at the double point will be projected as parallel asymptotes on the plane of projection. The figure will be a hyperbolism of the hyperbola without a diameter. Species 57, fig. 65.

5. Let D X be a tangent passing through the double point of the node, the projection will consist of two parabolic and a hyperbolic branch extending in one direction, and of a hyperbolic branch in the opposite direction. Species 66, or the trident, fig. 74.

6. Let D X cut the branches in three points. Species 2, fig. 8.

7. Let D X touch one and cut the other branch. The projection will consist of two hyperbolo-parabolic curves, one of which has a node. Species 47, fig. 55.

8. Let D X pass through the double point, and cut the folium, the species will be 58 or 59, according as the directrix bisects the folium or not. Figs. 66, 67.

Projections generated by the semi-cubic parabola.

Case 1. The three projections which are the result of the case in which the ordinates are parallel to the directrix, have already been noticed.

Case 2. When the ordinates of the given curve are oblique to the directrix.

1. Let D X cut the curve in one point, the projection will be a serpentine cusped hyperbola. Species 35, fig. 45.

2. Let it also pass through the cusp. The projection will be a hyperbolism of the parabola, without a diameter. Species 64, fig. 72.

3. Let it cut the curve in three points. Species 3, fig. 10.

4. Let it cut and touch the curve. Species 48, fig. 56.

5. Let the directrix coincide with the axis. The projection will be a cubic parabola. Species 72, fig. 81.

In the foregoing catalogue nearly all the Newtonian curves are accounted for, the exceptions being the 16th, 17th, 28th, and 29th species, all of which may be generated from the divergent parabola whose equation has two impossible roots, the curve which generates these projections is, however, not bell-shaped, but rather of the form by heralds called *nebuly*, the case in which this variety of parabola is produced has already been noticed.

Whenever it may be required from the given equation of any line of the third order, to find the equation of its projection, the inquiry will be assisted by recollecting that when the ordinates of the proposed curve are not parallel to the directrix, the projection will have no diameter, that whenever the directrix cuts the curve in three points, or in two points of which neither is a double point, the projection will be a redundant hyperbola, that whenever the directrix both cuts and touches the curve, the projection will be a hyperbolo-parabolic curve, that whenever the directrix cuts the curve in but one point, the generated projection will be a defective hyperbola intersecting its asymptote; finally, that whenever the directrix does not cut the curve at all, the projection will be a defective hyperbola having branches on the same side of their asymptote.

Clairaut in his treatise, "Sur les courbes que l'on forme en coupant une surface courbe quelconque par un plan donné de position," has treated this subject in a somewhat different manner. He supposes an immovable pole or luminous point, whence rays of light proceed, forming a cone of which the given curve is the base. And if one of the five divergent parabolas be assumed as a base, he shows that all the sections which a plane can make with the cone thus formed, will be curves of the same order.

To find all the different species which may be thus generated from the five cones of the third order, we must form the equation of the conical surface, and substituting in it the general equation to a plane,

$$x + my + nz = q,$$

thence deduce all the different curves which may be expressed by the transformed equations.

The subject has also been discussed in an elaborate paper by Nicole, in the Mémoires de l'Académie des Sciences, An 1731, under the title "Manière d'engendrer dans un corps solide toutes les lignes du troisième ordre." The conclusions at which he arrives are similar to those of Clairaut, although dependent on a different mode of reasoning.

Notes to Section VI.

" On the organic description of curves."

The demonstration of the first theorem in this section is given by Newton in the Principia, sect. v., lemma 21, and the subject is discussed further in the same section, prop. 22, &c. The reader may also with advantage consult Leslie's " Geometrical Analysis," Book II., Props. 5 to 12, where the description of conic sections by means of Newton's rotating angles is investigated at length. Or if he desires to know how the problem may be solved by the help of the anharmonic properties of conics, he will find the subject discussed in Mr. Salmon's " Treatise on Conic Sections," p. 281, 3d edition.

We may remark that the species of the conic section which is traced by the intersection of the legs A D, B D (fig. 82), will depend on the position of the given guiding line. If the angular points are joined by a straight line A B, and on it the segment of a circle is described such as may contain the supplement to 4 right angles of the given angles P A D, P B D.

Then if the guiding line cuts the circle twice, the conic section traced will be a hyperbola; and this hyperbola will be equilateral in the case of the given straight line passing through the centre of the circle. If the guiding line touches the circle, the result will be a parabola, if it falls altogether outside of the circle, an ellipse, which becomes a circle when the guiding line is infinitely distant.

Maclaurin has given a method by which curves may be traced, dependent on the same principle as that of the rotating angles, but which has the advantage of greater simplicity of construction. Instead of angles he makes use only of three straight lines made to revolve about fixed points as poles.

By this method a conic section may be described which shall pass through any five given points (no three of which lie in the same straight line). *See* Maclaurin's Algebra, and his papers in the Phil. Transactions. *See* also Leslie's " Geometry of Curves," Prop. 11.

We may observe that the number of the poles assumed is not

necessarily restricted to three, for if any number of poles be assumed having straight lines revolving on them, provided all the intersections but one are carried along given straight lines, that one will describe a conic section. The same observation holds good in the case of the rotation of angles about fixed poles, that is the curve described by the rotation of several angles, will not exceed a conic section, provided all the intersections of the legs except one, are carried along straight lines.

The Theorems II. and III. of this section are not of considerable importance in the investigation of the properties of curves of the higher orders, and the reasoning on which the demonstration of their truth depends, does not admit of condensation within the limits of a note. Newton has given no demonstration of them, but an elaborate exposition and illustration of the subject is to be found in the Geometria Organica of Maclaurin, who has also given two other methods for the description of these lines in the same work.

As regards lines of the third order, his methods may be generalised by the enunciation of the following theorem.

Two points in a plane C, and S, being given; if a straight line SN, making with the straight line NL a given angle SNL, revolve round S as a pole, and if the given angle FCO revolve about the pole C, then, if any two of the three points Q, P, N, in the quadrilateral CQPN are guided along straight lines given in position, the third will describe a line of the third order, having a double point in C or S.

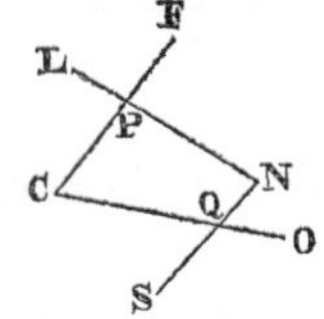

Note to Section VII.

" On the construction of equations by the description of curves."

" By the description of these curves, and their intersections, the roots of the equation to be constructed will be given."

Note to p. 29, line 15.

Suppose, for instance, it were required to find the roots of the equation of the fifth degree,

$$x^5 + a^2 x^3 - a^5 = 0 \qquad (1)$$

Assuming the parabola $x^2 = ay$, the new equation is,

$$xy^2 + axy - a^3 = 0 \qquad (2)$$

Let the curve indicated by this equation be described as shown in the figure, it will be a hyperbolism of the hyperbola referred to OX, OY as axes. Let the parabola $x^2 = ay$ be constructed to 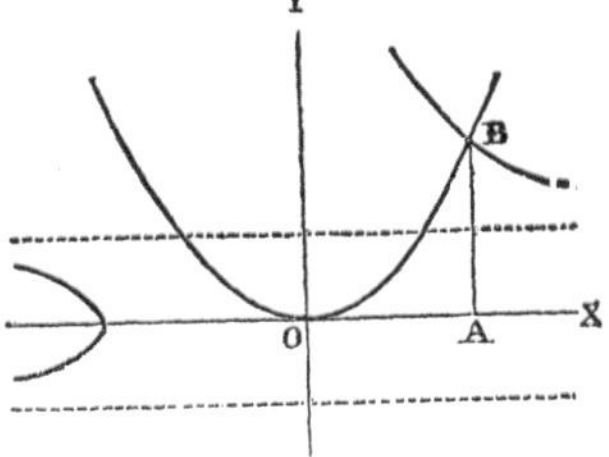 the same axes and origin, it will cut the branch of the hyperbolism in B. Then drawing the perpendicular AB, OA is the only real root belonging to the proposed equation (1).

In the same manner might the root be found by the assumption of the equation to the hyperbola $xy = a^2$, and the substitution of the value of x thus obtained in equation (1). In this case however the second equation becomes

$$x^3 + a^2 x - a y^2 = 0,$$

which indicates a semicubic parabola, and if this curve and also the hyperbola be described, as in the figure, the root is the same as before, viz., $OA = x$. 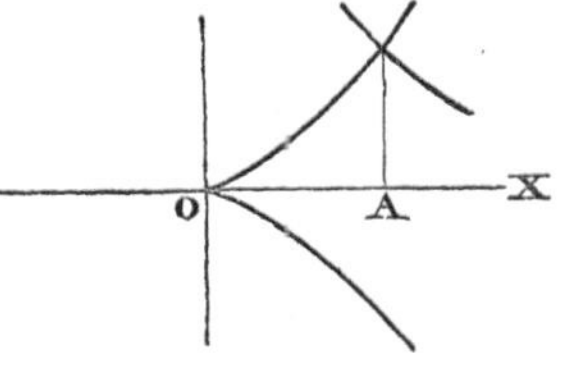

Our author remarks that there are many problems whose solution may be effected by the intersection of curve lines. Among these, is the celebrated Delian problem of the duplication of the cube, which depends on the finding of two mean proportionals between two given quantities, viz., the side of the given cube, and the side of its double; the first of the mean proportionals being the side of the double cube, as was observed by Hippocrates of Chios 2000 years ago.

The intersection of curve lines enables us to find any number of mean proportionals between two given quantities a, and b: for if x be the first of them they will form the progression,

$$a, \; x, \; \frac{x^2}{a}, \; \frac{x^3}{a^2}, \; \frac{x^4}{a^3}, \; \frac{x^5}{a^4}, \; \&c.,$$

If now *two* mean proportionals are wanted, the fourth term of the progression must be $= b$, and therefore we have,

$$\frac{x^3}{a^2} = b, \text{ or } x^3 = a^2 b, \text{ or } x^4 = a^2 b x,$$

by which the problem may be solved either by the intersection of two parabolas, by that of a parabola and hyperbola, or by that of a parabola and circle.

Let it be required to find *four* mean proportionals, in which case b should be the *sixth* term of the progression, and the equation to be solved is

$$x^5 - a^4 b = 0,$$

assuming the parabola $x^2 = a y$, the substitution gives for the locus,

$$x y^2 - a^2 b = 0,$$

which is that of the cubic hyperbola. This curve being described, will necessarily cut the parabola in one point P, and the absciss which corresponds to the ordinate of the point of intersection P, will be the root required, that is the first of the four mean proportionals, whence the others may be found.

The root might also have been obtained through the means of the hyperbola $x y = a^2$, by substituting in the original equation $x^2 y^2$ for a^4, giving for the locus

$$x^3 = a y^2,$$

a semicubic necessarily intersecting the hyperbola in one point.

If it be required to insert *six* mean proportionals between a and b, the intersection of two lines of the third order will solve the problem. Thus let x be the first of them,

$$\therefore \quad x^7 = a^6 b,$$

On the axes A C, A D with vertex A, describe the cubic parabola A G L, whose parameter A B = a, and the semicubic

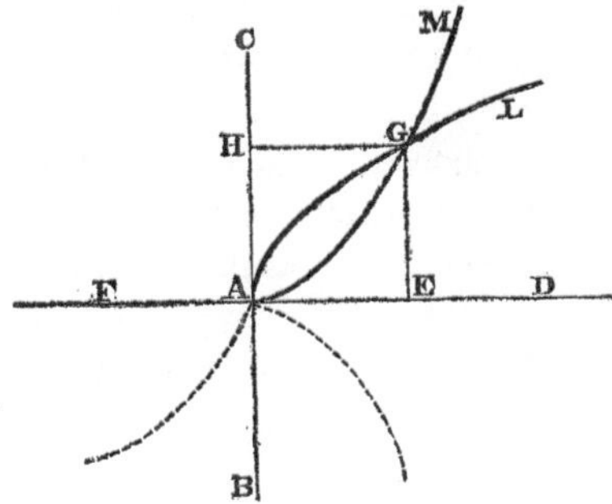

parabola AGM, whose parameter $AF = b$.   Let G be the point where the curves intersect, and draw the ordinates EG, HG.

In the cubic parabola, if $AH = EG = x$, $AE = HG = y$

$$AB^2 . AE = EG^3 \quad \therefore \quad y = \frac{x^3}{a^2} \quad \therefore \quad y^3 = \frac{x^9}{a^6},$$

In the semicubic, if $AE = HG = y$, $AH = EG = x$

$$AF . AH^2 = HG^3 \quad \therefore \quad y^3 = b x^2$$

$$\therefore \quad x^9 = a^6 b x^2, \quad \text{or} \quad x^7 = a^6 b$$

whence we have $EG = x =$ the first of the required means, and AE will represent the third, whence the rest may be found.

By the intersection of curve lines many other problems may be solved, but as it is evident that the correctness of the result must depend on the graphical accuracy with which the curve lines can be described, the method must be classed in the very numerous category of subjects which are more curious than useful.

On the Analytical Parallelogram.

Having brought to a conclusion the notes to the text of Sir I. Newton's enumeration of lines of the third order, we now propose to add some observations on the best method of distinguishing the species of any curve of which the equation is given. In ordinary practice it is usual to have recourse to the assistance of the differential calculus for this purpose, but another method founded upon the properties of the "analytical parallelogram," invented by Newton, may frequently be used with advantage, especially if the nature and direction of the asymptotes and infinite branches are the object of inquiry. It is indeed on this method that Cramer has founded most of the reasoning of his great work on curve lines, modestly entitled " Introduction à l'Analyse des Lignes Courbes," in which curves of great intricacy up to the sixth order are discussed, and in which not only the infinite parts of the curves, but their singular points and maximum and minimum co-ordinates are found, without any reference to principles beyond those of ordinary algebra.   The first notice of this invention appears in Newton's letter to Oldenburg, " de Seriebus condendis et invertendis," and was afterwards more fully explained in the Geometria Analytica. We proceed to consider this invention as applicable to the investigation of curve lines, although that was not the original object its inventor was aiming at when he first described it.

It frequently happens that the equation of a curve is presented in a form requiring tedious transformations in order to enable us to ascertain the species to which it belongs. Whenever such a difficulty occurs, the analytical parallelogram will be found a useful coadjutor. The peculiar value indeed of this ingenious invention appears to have escaped the notice of modern elementary writers, who when they mention it, usually content themselves with describing its construction without adverting to the assistance it is calculated to afford the student in the discussion of the properties of algebraic curves, particularly of those whose equations are not easy solvable. What the differential calculus is to the finite parts of the curve, the analytical parallelogram may be said to be with respect to the infinite portions of it, namely an important aid in the abridgment of laborious inquiry; not that it supplies any information absolutely unattainable by other means, but that it enables a much more speedy conclusion to be attained than could by other methods be made available.

The diagram annexed, which is taken from Sir I. Newton's " Geometria Analytica," is sufficiently explanatory of the con-

C

x^6	$x^6 y$	$x^6 y^2$	$x^6 y^3$	$x^6 y^4$	$a^6 y^5$	$x^6 y^6$	$x^6 y^7$
x^5	$x^5 y$	$x^5 y^2$	$x^5 y^3$	$x^5 y^4$	$x^5 y^5$	$x^5 y^6$	$x^5 y^7$
x^4	$x^4 y$	$x^4 y^2$	$x^4 y^3$	$x^4 y^4$	$x^4 y^5$	$x^4 y^6$	$x^4 y^7$
x^3	$x^3 y$	$x^3 y^2$	$x^3 y^3$	$x^3 y^4$	$x^3 y^5$	$x^3 y^6$	$x^3 y^7$
x^2	$x^2 y$	$x^2 y^2$	$x^2 y^3$	$x^2 y^4$	$x^2 y^5$	$x^2 y^6$	$x^2 y^7$
x	$x y$	$x y^2$	$x y^3$	$x y^4$	$x y^5$	$x y^6$	$x y^7$
1	y	y^2	y^3	y^4	y^5	y^6	y^7

A B

struction of the parallelogram. It will be seen that the terms are placed in the centre of each square and arranged in geometrical progression, not only in the vertical columns, as well as the horizontal, but also when taken in any straight line which passes through the centres of the cells or squares, for in any such cases the indices of x and y will be in arithmetical progression. *E. g.* In the terms y^7, $y^5 x$, $y^3 x^2$, $y x^3$, the indices of y decreasing by the common difference 2, while the indices of x increase in the progression of the natural numbers, the common ratio of the terms is $\dfrac{x}{y^2}$. It follows, that if any two terms be supposed equal, then all the terms in the same straight line with these terms will be equal, because, by supposing these two terms equal, the common ratio is supposed to be a ratio of equality, and, therefore, if we substitute for x, its value expressed in terms of y, the dimensions of y in all the terms *in* the straight line will be *equal*, while the dimensions of y in the terms *above* the straight line will be *greater*, the dimensions of y in the terms *below* the straight line will be *less* than its dimensions in that line. *E. g.* by supposing $y^7 = y x^3$, we find $y^6 = x^3$, whence $x = y^2$. Substituting therefore y^2 for x, the dimensions of y in the terms y^7, $y^5 x$, $y^3 x^2$, $y x^3$ will be 7, but the dimensions in all the terms above that line will be greater than, in all the terms below that line less than, 7.

Hence it follows that if we write all the terms of an equation $f(x y) = 0$ in their proper squares, and form a convex polygon by lines joining these terms, those sides of the polygon which are uppermost when the parallelogram is placed with A B horizontal will represent the equation when x is assumed infinitely great, those sides which are undermost representing the equation when x is assumed infinitely small. And if we desire to know the consequences of assuming y infinitely great or small, we have only to place the parallelogram with the side A C horizontal, and then to draw similar inferences. For if in an indeterminate equation, x, or y, is supposed infinite, the supposition renders certain terms of the equation infinitely greater than the remaining terms, which latter may, therefore, be neglected, and thus the greater terms will constitute the whole equation. The polygon enables us to ascertain which are the terms to be selected as being effective for the required purpose, and which of them may be neglected as irrelevant to the object in view.

We may further observe, that when there are several upper or several lower sides to the polygon, each of such sides will have its peculiar significance, and by forming separate equations from each of them several results will be obtained for the same curve.

But as the use of the parallelogram is better shown by examples than by description, we shall proceed to give some illustrative cases.

Suppose the equation is $x^2y + ay^2 - a^2x = 0$,

and it is required to ascertain the species to which this equation belongs. Let the terms be marked on their proper squares on the parallelogram, as shown in the figure: if we desire to know the result of the hypothesis that $x = \infty$, let the diagram be placed 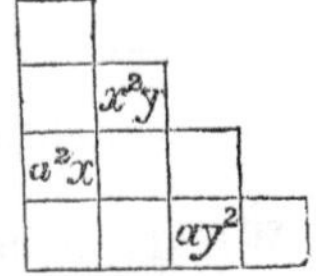 with the column containing only y's horizontal. The polygon will then show two upper sides, from which we form the equations

$$x^2y - a^2x = 0, \text{ or } xy - a^2 = 0 \qquad (1)$$
$$x^2y + ay^2 = 0, \text{ or } x^2 + ay = 0 \qquad (2)$$

From (1) we obtain the hyperbolic branches which form the curve at infinity, positive and negative.

(2.) No branch in the direction of the axis of x.

Let the parallelogram be now placed with the row of x's horizontal, and the polygon then shows only one upper line, showing the consequence of y infinite, viz.,

$$x^2y + ay^2 = 0, \text{ or } x^2 + ay = 0,$$

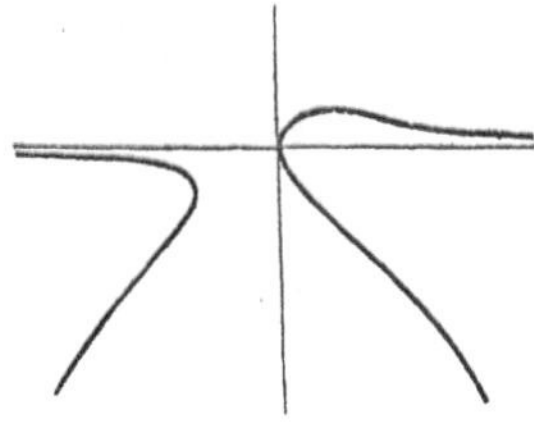

which, when y is negative, indicates a parabola, the curvilinear asymptote to which the branch approaches. We therefore infer the curve to belong to one of the seven parabolic hyperbolas. Its figure will be as shown.

Example 2. Let

$$xy^2 - a^2y - b^3 = 0.$$

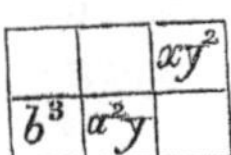

Placing these terms in their squares as before, there are two equations obtainable, one for x infinite and one for y infinite, viz.,

$xy^2 - b^3 = 0$, representing the branches of the cubic hyperbola when x is infinite,

$xy^2 - a^2y = 0$, or $xy - a^2 = 0$, representing hyperbolic branches when y is infinite.

The curve is, therefore, a hyperbolism of the parabola without a diameter, Species 64.

Example 3.   Let

$$xy^2 - 12y + x^3 - 10x^2 + 25x = 0.$$

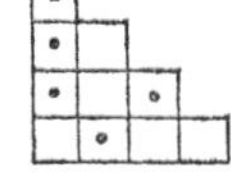

When x is infinite there is only one upper side to the polygon, viz.,

$$xy^2 + x^3 = 0 \qquad (1)$$

shewing only imaginary branches.

When y is infinite, in addition to (1),

$$xy^2 - 12y = 0, \text{ or } xy - 12 = 0 \qquad (2)$$

The curve is evidently a defective hyperbola, species 33 of the enumeration.

Example 4.   Let

$$xy^2 - ay^2 - 3axy + 2a^2x = 0.$$

x being supposed infinite we have

$$xy^2 - 3axy + 2a^2x = 0, \text{ or } y^2 - 3ay + 2a^2 = 0,$$

whence $y = a$, and $2a$. (1)

For y infinite the equation is $xy^2 - ay^2 = 0$, whence $x = a$. (2)

Equation (1) marks the existence of the two parallel rectilinear asymptotes corresponding to the ordinates a and $2a$. Equation (2), $x = a$, likewise indicates an asymptote parallel to the ordinates at a distance $= a$ from the origin. The curve is a hyperbolism of the hyperbola.

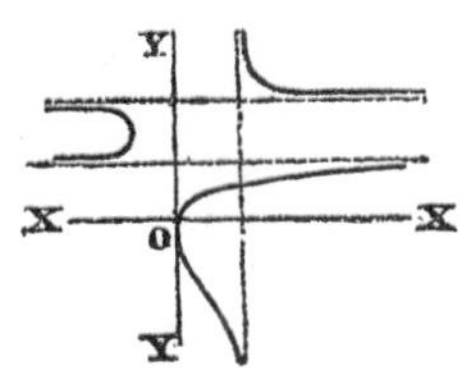

In general, whenever the curve admits of infinite branches and the side of the polygon is parallel to one of the sides of the parallelogram, we may infer the existence of rectilinear asymptotes parallel to one or other of the axes.

Example 5.   Let

$$x^2 y^2 - a^3 y - b^3 x = 0.$$

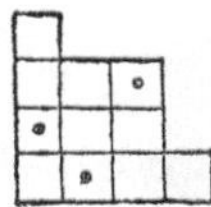

The upper side when x is made infinite gives

$$x^2 y^2 - b^3 x = 0, \text{ or } x y^2 = b^3,$$

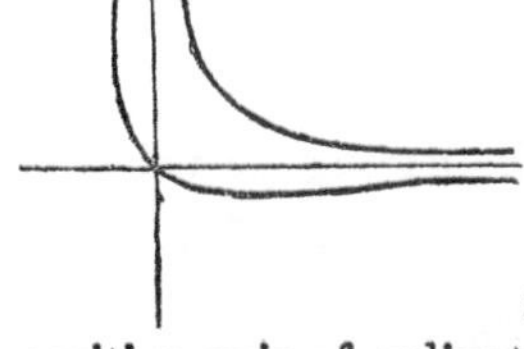

and when y is made infinite

$$x^2 y^2 - a^3 y = 0, \text{ or } x^2 y = a^3,$$

there are therefore four infinite branches, one on each side of the positive axis of abscisses, and one on each side of the positive axis of ordinates.

Example 6.   Let the equation be

$$y^4 + 2 x^2 y^2 + x^4 - 6 a x y^2 - 2 a x^3 + a^2 x^2 = 0,$$

x being infinite, the equation $y^4 + 2 x^2 y^2 + x^4 = 0$ gives only imaginary results; and the same remark being true for y infinite, it results that the curve has no infinite branch.   See figure at page 46.

In order more fully to exemplify the power of the analytical parallelogram, a few examples are here added, involving equations of higher dimensions.

Example 7.   In the equation

$$y^4 x - y^3 x^2 + 3 y x^3 - y^2 x + 4 y - 2 x = 0,$$

it is required to find the number and nature of the infinite branches.   First, assuming x infinite,

$$(1) \quad y^4 x - y^3 x^2 = 0, \text{ or } y = x$$

shewing two infinite branches accompanying an asymptote cutting the axes at an angle of 45°.

$$(2) \quad 3yx^3 - y^3x^2 = 0, \quad \text{or } 3x = y^2 \therefore y = \pm \sqrt{3x}$$

two parabolic branches, above and below the axis of x, on the positive side.

$$(3) \quad 3yx^3 - 2x = 0, \quad \text{or } 3yx^2 - 2 = 0, \therefore y = \frac{2}{3x^2}$$

two hyperbolic branches of the third order.

Next assuming y to be infinite, in addition to the equations (1) and (2), we have

$$y^4x + 4y = 0, \quad \text{or } y^3x + 4 = 0 \therefore x = -\frac{4}{y^3}$$

whence the axis of y is itself an asymptote, having branches of the curve on each side, for the term $-\dfrac{4}{y^3}$ is negative for y positive, and positive for y negative. There are therefore eight infinite branches.

Example 8.   Let

$$y^5x^3 - y^4x^6 + y^2x^5 - 4y^3x^3 + 2y^4x + x^4y - 5y^2 + axy - 3x^2 = 0.$$

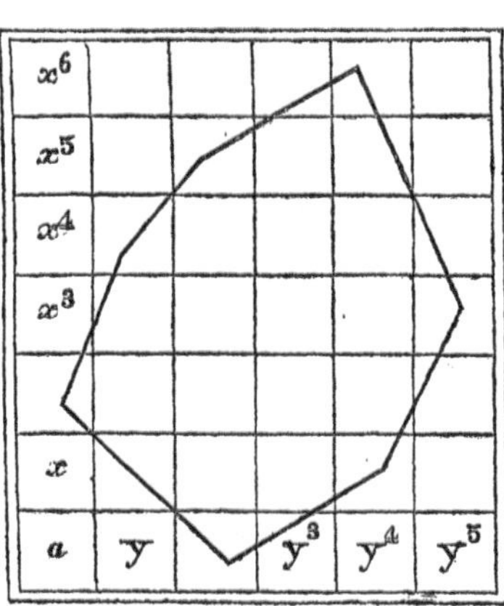

The equations obtained when x is supposed infinite here, are

$$(1) \quad y^5 x^3 - y^4 x^6 = 0, \quad \text{or } y = x^3$$

$$(2) \quad y^2 x^5 - y^4 x^6 = 0, \quad \text{or } xy^2 = 1, \quad \therefore \ y = \pm \frac{1}{\sqrt{x}}$$

$$(3) \quad y^2 x^5 + x^4 y = 0, \quad \text{or } y = \frac{1}{x}$$

$$(4) \quad x^4 y - 3 x^2 = 0, \quad \text{or } y = \frac{3}{x^2}$$

When y is assumed infinite, besides (1),

$$(5) \quad y^5 x^3 + 2 y^4 x = 0, \quad \text{or } x = \pm \sqrt{-\frac{2}{y}}$$

$$(6) \quad 2 y^4 x - 5 y^2 = 0, \quad \text{or } x = \frac{5}{2 y^2}$$

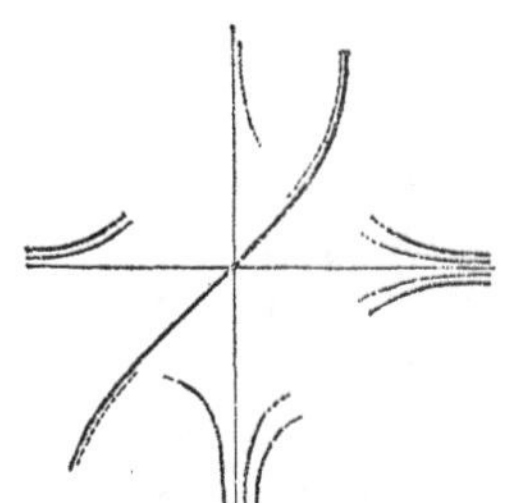

(1) Indicates a cubic parabola, the asymptote of two branches of the curve.

(2) A hyperbolic asymptote furnishing two branches on the positive side.

(3) Positive and negative branches.

(4) Two hyperbolic branches converging more closely to the axis than those of (2) or (3).

(5) Two hyperbolic branches in the negative region of y.

(6) Two hyperbolic branches.

The curve has therefore apparently twelve infinite branches.

Thus far we have not remarked upon the significance of the lower sides of the polygon, which give the consequences of x or y infinitely small, on the same principle that the upper sides shew the result of x or y infinitely great. However ingenious is the theory by which the singular points, maxima and minima ordinates, &c. of curves, are inferred from the equations obtainable from the lower sides of the polygon, yet, since the differential calculus is in general better adapted for the discussion of infinitesimal quantities, we shall not notice this use of the parallelogram, but proceed to advert to the case in which neither x nor y is supposed to be infinite.

The parallelogram is now supposed to be placed with its

angle downwards, and in that position it is evident that each row of terms read horizontally is composed of terms of the same dimensions. The lowest cell is occupied by a constant, viz., the absolute term; the first row consists of the terms of a simple equation, the second row consists of the terms of a quadratic, the third row of the terms of a cubic, &c. This disposition of the terms furnishes several useful hints for the formation of rules for ascertaining the nature of the curve near the origin. Thus, when the lowest cell is vacant, we see that the curve passes through the origin; when the first row also is vacant, there is a double point at the origin; when the second or quadratic row is also vacant, there is a triple point at the origin; when the third row also, a quadruple point; and so on.

Again, to ascertain the direction of the curve at the origin, that is, to draw a tangent there, we have only to equate to zero the lowest row, and construct the straight line or lines which this equation indicates. Such straight line or lines will be tangent to the curve at the origin. When the equation so determining the tangent shows that $y=0$, the axis of x touches the curve; if the roots of the tangential equation are imaginary, the series derivable therefrom are also imaginary, and in this case the origin is a conjugate point; and if the roots of the equation are equal, there is a double point at the origin, formed by the contact of two branches having a common tangent, or rather the coincidence of two tangents, indicating (where the line is of the third order) a cusp.

For example, take the equation

$$y^4 - 2y^2x^2 + x^4 + 2ay^2x - 5ax^3 = 0$$

The first and second rows are vacant when this equation is referred to the analytical parallelogram; there is, therefore, a triple point on the origin. The lowest row is

$$2ay^2x - 5ax^3,$$

which equated to 0 gives

$$x=0, \quad y\sqrt{2} - x\sqrt{5} = 0, \quad y\sqrt{2} + x\sqrt{5} = 0$$

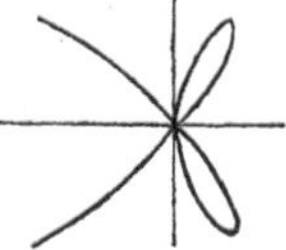

There are, therefore, three tangents, one of which is the axis of y, and the other two are determined by taking on the absciss $\sqrt{2}$, the ordinates $\pm \sqrt{5}$, and drawing straight lines from the origin to the extremities of the ordinates.

If the tangential equation is such as that it will divide one of the rows of terms superior to it, another inference may be drawn, viz., that the origin is a point of inflexion, and of a degree dependent upon the number of superior rows which each root of the tangential equation will divide,

Example. For the curve denoted by $x^3 - a x y - b^2 y = 0$, the tangential equation is $-b^2 y = 0$, or $y = 0$

This root, substituted in the second row, makes it also $= 0$, whence the origin is a point of inflexion; but the third row is not divisible by $y = 0$, therefore the point of inflexion is simple.

Example 2. In the equation

$$y^4 + 2 y^2 x^2 - 4 a y^3 - 4 a x^2 y + 8 a^2 y^2 - 8 a^3 y = 0$$

(Cramer, *Méthode des Tangentes*, § 185.)

The first row $-8 a^3 y$, gives $y = 0$, which value, substituted in the superior rows, makes the second and third disappear, but not the fourth. The origin is, therefore, a point of double inflexion, or *serpentement*.

If the point in question be not the origin, it is always possible to transport it thither by a transformation; and then, by a mere inspection of the analytic parallelogram, we can ascertain the simplicity or multiplicity of any assigned point, the direction of its tangent and the degree of its inflexion. This use of the Newtonian invention was first pointed out by the Abbé de Gua, in his *Usage d'Analyse de Descartes*.

Another use which may be made of the Newtonian parallelogram consists in the development of equations by infinite series to be effected by its means. This, indeed, was the object which the illustrious inventor had in view when he originally designed it, as may be seen in the explanation of its use to be found in his letter to Oldenburgh, and in the *Geometria Analytica*. This use may be made available, as being supplemental to that which we have been discussing, for the investigation of curves, we shall therefore here shortly notice it.

An algebraic equation in x and y being given, to find the value of y in terms of x, expressed in a descending series, the parallelogram is to be placed with the row without x horizontal, and the first term of the series is to be obtained as has been already shown, the second and subsequent terms being obtained as is shown in the following example. And to find the first term of an *ascending* series, the equations must be formed from the lower sides of the polygon. Similarly, to obtain a series giving the value of x in terms of y, it is only necessary to place the row without y horizontal, and then proceed as before shown.

For example, in the equation

$$y^2 - 2xy + x^2 - 2ay + ax + a^2 = 0 \qquad (1)$$

to find the value of y in terms of x in an ascending series, the lower side of the triangle gives but one equation, $y^2 - 2ay + a^2 = 0$, whence $y = a$, the first term of the series.

To find the second term, substitute this value $+ u$, and apply the transformed equation to the analytical triangle arranged in powers of x and u

$$\therefore \ u^2 - ax - 2ux + x^2 = 0 \qquad (2)$$

and this, applied to the triangle, gives one lower side, viz. $ax = u^2$, whence $u = \pm a^{\frac{1}{2}} x^{\frac{1}{2}}$, which is the second term of the series.

To find the third term of the series, substitute $\pm a^{\frac{1}{2}} x^{\frac{1}{2}} + t$ for u in (2), and there results

$$\pm 2 a^{\frac{1}{2}} t x^{\frac{1}{2}} + t^2 \pm 2 a^{\frac{1}{2}} x^{\frac{3}{2}} - 2tx + x^2 = 0 \qquad (3)$$

Before applying this equation to the triangle it is necessary to insert the terms in which the exponent of x is $\dfrac{1}{2}$, midway between the squares x and tx, and that in which the exponent of x is $\dfrac{3}{2}$ midway between the squares containing x^2 and x.

The lower side will give $\pm 2 a^{\frac{1}{2}} t x^{\frac{1}{2}} \pm 2 a^{\frac{1}{2}} x^{\frac{3}{2}} = 0$, or $t = x$, which is the third term of the series.

If to find the fourth term of the series we substitute $x + s$ for t in the equation (3) we only obtain an exponent for s whose value is less than that of the preceding one. There is therefore no fourth term, and the series is terminated, being

$$y = a \pm \sqrt{ax} + x$$

Again the equation

$$x^2 y + a y^2 - 2 a x y + a x^2 = 0,$$

following the same process, furnishes the series,

$$y = x \pm \sqrt{-\frac{x^3}{a} - \frac{x^2}{2a}} \cdots$$

possible only when x is negative.

And the equation

$$x^3 + x^2 y + a y^2 - 2 a^2 y + a^2 = 0,$$

furnishes a series whose second term is imaginary, showing that it is not possible to express y in terms of x by an ascending series.

If the equation be

$$y^3 + a^2 y - 2 a^3 + a x y - x^3 = 0,$$

the series will be

$$y = a - \frac{x}{4} + \frac{x^2}{64a} + \frac{131 x^3}{512 a^2} + \&c.$$

which is the example given by Newton.

When it is desired to express x in terms of y by a *descending* series, the process for obtaining the first and subsequent terms of the series required, is precisely the same as that for obtaining the ascending series, with the exception that the upper sides of the analytical polygon are to be chosen, wherefrom to form the separate equations.

Thus in example 7, from equation (1), the first term of a series descending by powers of x, and therefore converging the faster in proportion as x is assumed greater, was found to be $y = x$.

To find the second term of the series let $x + u$ be substituted for y, where u is supposed to represent all the remaining terms of the series, and the transformed equation being placed on the parallelogram, gives one effective upper side, viz.

$$u x^4 + 3 x^4 = 0, \text{ whence } u = -3$$

which is the second term of the series.

Then assuming $u = -3 + t$ and again applying the transformed equation as before, and assuming x infinite, we obtain the third term of the series, which proceeds

$$y = x - 3 - \frac{27}{x} \dots \text{ \&c.}$$

whence the equation to the rectilinear asymptote is $y = x - 3$, the negative sign of the third term shows that the branch accompanying this asymptote is to be found on the lower side of this line while x is positive, and the upper side when x is negative.

In the same example from equation (3) the first term of the descending series is $y = \pm 3^{\frac{1}{3}} x^{\frac{1}{3}}$, and proceeding in the same manner as in the last case, the second term is found to be $+ \frac{3}{4}$. If another term of the series be calculated, the term $x^{\frac{1}{3}}$ appears in the denominator. The series therefore shows that the branches approach a parabolic asymptote whose axis is the line $y = \frac{3}{4}$.

So again in example 8, the first term of a descending series being found by equation (1), viz. $y = x^3$, substituting in the equation $x^3 + u$ for y, the second term is found to be $- \frac{2}{x^2}$, whence it appears that the two branches of the curve converge to the branches of a cubic parabola, that in the positive region below the asymptote, that in the negative region, above it.

The following rule will be found to abridge the labour necessary to adapt the transformed equations to the analytical parallelogram.*

Multiply each term of the equation which contains y by the exponent of y in such term, and also by $A x^h$ the ascertained value of y when x is supposed infinite, and divide such term by y. A

* Cramer, *Analyse des Lignes Courbes.*

new row of terms being thus obtained, multiply each of them by *half* the exponent of y, by $A x^h$, and divide by y. A third row is to be subjected to the same process, only that $\dfrac{1}{3}$ the exponent of y is to be used, and so on until y no longer appears in the remaining terms. Collecting these terms and referring them to the analytical parallelogram, one of the upper sides will indicate the equation which gives the second term of the series, and this new value of y is to be used for finding the third term, the process being repeated according to the accuracy required. It is necessary to remember that here by y is intended " the remainder of the series."

For example, in the first case of the Newtonian equations,

$$xy^2 + ey - ax^3 - bx^2 - cx - d = 0$$

The first term of the series being $\pm \sqrt{ax}$, $A = \pm \sqrt{a}$, and $h = 1$, then the process is as follows : —

$$xy^2 + ey - ax^3 - bx^2 - cx - d = 0$$
$$2 \qquad 1 \qquad 0 \qquad 0 \qquad 0 \qquad 0$$

$$\pm 2 \sqrt{a}\,x^2 y \pm e \sqrt{a}\,x$$
$$\tfrac{1}{2} \qquad\qquad 0$$

$$ax^3$$

Collecting the terms, the equation is

$$xy^2 + ey - bx^2 - (c \pm e \sqrt{a})\, x \pm 2 \sqrt{a}\,x^2 y = 0$$

whence $y = \pm \dfrac{b}{2 \sqrt{a}}$, the second term of the series.

$$xy^2 + ey - bx^2 - (c \pm e \sqrt{a})\, x \pm 2 \sqrt{a}\,x^2 y$$
$$2 \qquad 1 \qquad 0 \qquad\qquad 0 \qquad\qquad 1$$

$$\pm \dfrac{bxy}{\sqrt{a}} \pm \dfrac{be}{2 \sqrt{a}} \qquad\qquad\qquad + bx^2$$
$$\tfrac{1}{2} \qquad\quad 0 \qquad\qquad\qquad\qquad\qquad 0$$

$$\dfrac{b^2 x}{4a}$$

Collecting the terms, the equation is

$$x y^2 \pm 2 \sqrt{a}\, a x^2 y + \left(\frac{b^2}{4a} - c \pm e \sqrt{a} \right) x = 0$$

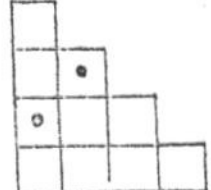

whence $y = \dfrac{b^2 - 4 a c \pm 4 a e \sqrt{a}}{8 a x \sqrt{a}}$, the third term of the series.

Example 2.   Let $x^3 - a x y + y^3 = 0$

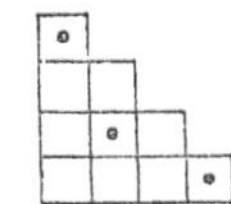

Here the first term is $-x$, $\therefore$ A $= -1$, and $h = 1$,

$$
\begin{array}{ccc}
x^3 & - a x y & + y^3 \\
0 & 1 & 3 \\
\hline
& a x^2 & - 3 y^2 x \\
& 0 & 1 \\
\hline
& & 3 y x^2 \\
& & \tfrac{1}{3} \\
\hline
& & - x^3 \\
\end{array}
$$

Collecting terms, the equation becomes

$$y^3 - a x y + a x^2 - 3 y^2 x + 3 y x^2 = 0$$

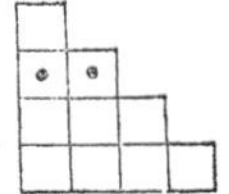

whence $y = -\dfrac{a}{3}$, the second term of the series.

$$
\begin{array}{ccccc}
y^3 & - a x y & + a x^2 & - 3 y^2 x & + 3 y x^2 \\
3 & 1 & 0 & 2 & 1 \\
\hline
a y^2 - \dfrac{a^2}{3} x & & & + 2 a x y & - a x^2 \\
1 & 0 & & \tfrac{1}{2} & 0 \\
\hline
- \dfrac{a^2 y}{3} & & & - \dfrac{a^2 x}{3} & \\
\tfrac{1}{3} & & & 0 & \\
\hline
\dfrac{a^3}{27} & & & & \\
\end{array}
$$

Collecting terms, the equation gives for the highest terms,

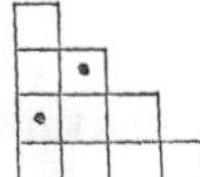

$$3\,y\,x^2 - \frac{2\,a^2\,x}{3} = 0$$

whence $y = \dfrac{2\,a^2}{9\,x}$, the third term of the series.

Therefore, $\qquad y = -\,x - \dfrac{a}{3} + \dfrac{2\,a^2}{9\,x} \ \ldots \ldots \ \&\text{c.}$

The reader who may desire further information relative to the important theorems deducible from the Newtonian parallelogram, will find the subject discussed at great length in Stewart's "Analysis by Equations Explained," published in 1745; and he is further referred to a paper on "Newton's Method of Co-ordinated Exponents," by Professor De Morgan, in the Transactions of the Cambridge Philosophical Society, vol. ix. part 4, as well as to the work of Cramer already cited.

The method given above of finding the value of one co-ordinate in terms of the other by a descending series, is useful for determining the asymptotes of curves, since the first two terms are generally all that need be computed. Previously to making use of this or any of the rules derived from the analytical parallelogram, it is desirable to notice whether the given equation is resolvable into factors, for in such a case it will represent only a combination of lines of an inferior order.

It is also necessary to observe whether the sum of the indices in the equation be the same in every term, for then lines joining the centres of the squares containing these terms on the analytical parallelogram, will not form a polygon or triangle, but will lie in one straight line. In this case the locus consists of only straight lines. Thus if

$$y^3 - 2\,x\,y^2 + x^3 = 0$$

dividing by x^3,

$$\left(\frac{y}{x}\right)^3 - 2\left(\frac{y}{x}\right)^2 + 1 = 0$$

The roots of which are $1,\ \dfrac{1 + \sqrt{5}}{2},\ \dfrac{1 - \sqrt{5}}{2}$

therefore $y = x$, $y = \dfrac{1 + \sqrt{5}}{2}\, x$, $y = \dfrac{1 - \sqrt{5}}{2}\, x$

showing three straight lines passing through the origin, making the respective angles indicated with the axis of x.

Examples.

The examples which follow are intended to illustrate the application of the foregoing rules, to the discussion of lines of the third order, whose equations are given.

Ex. 1.   The equation to the curve is

$$y^3 - x^3 - a^3 = 0$$

Referring these terms to the analytical parallelogram we obtain for x infinite,

$$y = x + \frac{a^3}{3\,x^2} + \ \ldots \ \&c.$$

The asymptote therefore passes through the origin and cuts the axis at an angle of $45°$. The branches always lie above the asymptote, since the second term of the series is always positive. The curve has therefore a diameter. There being no multiple point indicated by the equation, and two of the roots being imaginary, the curve can be no other than the 45th of the enumeration.

Ex. 2.   The equation is $x y^2 + y - x^3 = 0$

The parallelogram here gives one equation for x infinite and one for y infinite, viz.

$$x y^2 - x^3 = 0, \quad \text{or } y^2 = x^2 \qquad (1)$$

$$x y^2 + y = 0, \quad \text{or } x y = -1 \qquad (2)$$

The series obtainable from (1) are

$$y = x - \frac{1}{2x} \ \&\text{c.} \quad y = -x - \frac{1}{2x} \ \&\text{c.}$$

showing two rectilinear asymptotes intersecting each other at right angles at the origin. From (2) the axis of y is itself an asymptote.

Solved for y, the equation gives

$$y = -\frac{1}{2x} \pm \sqrt{x^2 + \frac{1}{4x^2}}, \text{ and for } x \text{ negative, } y = \frac{1}{2x} \pm \sqrt{x^2 + \frac{1}{4x^2}}$$

the curve is therefore similar on each side of the origin.

When $x = 0$, $y = 0$ and $\pm \infty$. There are two real ordinates for every value of the absciss positive or negative. The curve will lie in the first and fourth quadrant below the asymptote, and in the second and third above it. There is a point of inflexion on the origin, and the four branches which do not pass through the origin are inscribed.

The species is No. 27 of the enumeration.

Ex. 3. The equation is $xy^2 + x^2y - a^3 = 0$

The axis of x is an asymptote, as also is the axis of y, the infinite branches being in each case on the positive side of the axis. There is a third asymptote passing through the origin, having branches below it. The equation shows there is no multiple point or point of inflexion. There are three diameters.

The species is No. 32 of the enumeration.

Ex. 4. The equation is $ay^2 - bx^2 - x^3 = 0$

There is one pair of parabolic branches. The origin is a double point, which as the curve is possible on each side of it is a point of intersection. When $x = -b$, $y = 0$, therefore there is a node on the negative side of the origin. There are maximum ordinates when $x = -\frac{2b}{3}$. The equation giving the tangents at the point of intersection is

$$y^2 = \frac{b}{a} x^2, \text{ and } \therefore \begin{cases} y = x\sqrt{\dfrac{b}{a}} + \dfrac{x^2}{2\sqrt{ab}} + \&\text{c.} \\[2ex] y = -x\sqrt{\dfrac{b}{a}} - \dfrac{x^2}{2\sqrt{ab}} \&\text{c.} \end{cases}$$

The first term gives the direction of the branches, and the second shows they are convex to the axis.

The curve is a nodate parabola.   Species 68.

Ex. 5.   Trace the curve whose equation is

$$4y^3 - 6xy^2 + 2x^3 + 2ay^2 + 4ax^2 - b^3 = 0$$

Applied to the analytical parallelogram, whether for x or y infinite, the resulting equation is

$$4y^3 - 6xy^2 + 2x^3 = 0$$

which is resolvable into the factors

$$4y + 2x = 0, \text{ and } y^2 - 2xy + x^2 = 0$$

whence $$y = -\frac{x}{2}, \; y = x, \; y = x,$$

are the first terms of the series, showing the infinite branches. Calculating the second and third terms of these series according to the rule given, p. 100.

$$y = -\frac{x}{2} - \frac{a}{2} + \frac{2a^2}{9x}, \text{\&c.} \qquad (1)$$

$$y = x \pm \sqrt{-ax} \pm \frac{a^2}{6\sqrt{-ax}}, \text{\&c.} \quad (2)$$

From (1) we see that an asymptote cuts the axis of x at an angle $\tan^{-1}\left(-\frac{1}{2}\right)$, and at a distance $-a$ from the origin. The uneven power of x in the denominator of the third term shews that on the positive side the curve will be above, on the negative side, below the asymptote.

From (2) we infer that the parabolic branches indicated exist only in the negative region, for the series is imaginary when x is positive. The branches are outside the parabolic asymptote. The equation $y = x \pm \sqrt{-ax}$, shews that the diameters of this parabola are inclined to the axis of x at an angle of 45°.

The curve is a parabolic hyperbola without a diameter. Species 50.

Ex. 6.   The equation is

$$y^3 - xy^2 + x^2y - x^3 + ay^2 - 2\,axy + 3\,ax^2 - 2\,a^2x = 0$$

The first term of the series obtained cn the hypothesis that x is infinite, is $y = x$, and proceeding according to the rule given at page 100, the second term is found to be $-a$, but the third term is either 0 or imaginary, so that when x is $\pm \infty$ the locus is the straight line

$$y = x - a$$

There is no branch of a curve either for x or y infinite, and therefore the equation refers to no curve belonging to the enumeration. Between 0 and $2\,a$ there are three values for the ordinate to any assumed absciss, by calculating one or two of which it is easy to see that the remainder of the locus consists of the circle touching the axis of y at the origin, and having the radius a.

Ex. 7.   To compare the curves expressed by the equations

$$y^2\,(x - a) = x^2\,(x + a) \quad . \quad . \quad . \quad (1)$$
$$y^2\,(a - x) = x^2\,(x + a) \quad . \quad . \quad . \quad (2)$$

Expanding for the asymptotes in (1) the result is

$$y = \pm \left(x + a + \frac{a^2}{2\,x} + \&c.\right)$$

also y is infinite when $x = a$. There are therefore three asymptotes, of which two intersect each other, and the third is parallel to the axis of y. If the equation is referred to the analytical parallelogram with the angle downwards, the first two rows of terms are vacant, therefore the origin is a double point, which in this case is obviously a conjugate point, for y is imaginary on each side of it. The species is the 13th of the enumeration.

For the equation to determine the asymptotes in (2) we obtain

$$xy^2 + x^3 = 0, \text{ or } y^2 + x^2 = 0$$

showing that the two asymptotes which were real in (1) are imaginary in (2), but the third asymptote indicated by $x = a$ still remains.

The origin is in this curve, as in the former one, a double point, but it is now a point of intersection, since y is possible on each side of it. The species is No. 41 of the enumeration.

Ex. 8.

$$x^3 + 10\,x^2 + x - y^2 + 10 = 0$$

The asymptote is here a semicubic parabola. Since this equation has but one real root, there is no oval node or conjugate point. When x is positive the values of $\pm\, y$ increase to infinity; when x is negative and greater than 10, y is imaginary. Between the origin and the vertex there are both maxima and minima values for y; the figure of the branches is represented at page 69.

Ex. 9.

$$xy^2 - x^3 + x^2 + 8\,x - 12 = 0$$

There is a double point when $x = 2$, the curve has six infinite branches, and is a cruciform hyperbola. Species 18, fig. 30.

Ex. 10.

$$xy^2 + x^3 + 9\,x^2 - 12\,x - 20 = 0$$

For x infinite, the branches are imaginary. For y infinite there are two hyperbolic branches to the axis of y as asymptote. The absciss is a diameter; when x is negative there is a conjugate oval.

Species 40, fig. 49.

Ex. 11.

$$y^3 - x^3 + 9\,x^2 - 18\,x = 0$$

The asymptote cuts the axis at the point $x = 3$, making an angle of 45° with it. There are three points of inflexion, all in the axis of x, viz., $x = 0$, $x = 3$, $x = 6$. There is a general centre at the point $x = 3$, for the even power of x disappears when the origin is transported thither. The axis of y is a tangent at the origin. The curve is a defective hyperbola with a centre, but no diameter. Species 38.

Ex. 12. The curve is

$$y^3 - 2xy^2 + x^2 y - a^3 = 0$$

When x is infinite there are two equations derivable,

$$x^2 y - a^3 = 0 \quad (1) \quad \text{and,} \quad y^3 - 2xy^2 + x^2 y = 0 \quad (2)$$

From (1) we infer that the absciss is an asymptote, for the series for y is

$$y = \frac{a^3}{x^2} + \&\text{c.,}$$

the curve being always above the asymptote. From (2) we infer that $y^2 - 2xy + x^2 = 0$, the factors of which are $x - y$, $x - y$. There is therefore a double asymptote cutting the axis at an angle of 45°. The series obtainable from (2) are $y = x \pm \dfrac{a^{\frac{3}{2}}}{x^{\frac{1}{2}}} - \&\text{c.,}$ and therefore the curve exists above and below the double asymptote, but only in the positive region of ordinates.

Since the curve does not intersect either asymptote, it has a diameter, and therefore the species is the 65th. The double asymptote is, in this case, the diameter belonging to the ordinates drawn parallel to the axis of x.

Ex. 13. The curve is

$$xy^2 - x^2 + x - 4y + 4 = 0$$

The series for y commence with

$$y = \frac{4}{x}, \text{ and } y = \pm \sqrt{x}$$

There are therefore hyperbolic as well as parabolic branches. Solved for y, the equation gives $y = \dfrac{2}{x}$ for the curvilinear diameter bisecting chords parallel to the axis of y, and solved for x, the parabola $2x = y^2 + 1$, is the curvilinear diameter to chords parallel to the axis of x.

The curve is species 52 of the enumeration.

Ex. 14.

$$xy^2 - x^2 - 3x - 2y - 3 = 0$$

The curvilinear diameters are the hyperbole $xy = 1$, and the parabola $y^2 = 2x + 3$. There is a cusp at the point $x = -1$, $y = -1$. The species is numbered 48, fig. 56.

Ex. 15.

$$xy^2 - x^2 - 6x - 4y - 9 = 0$$

The curve is a parabolic hyperbola, species 47. There is a point of intersection when $x = -1$, $y = -2$, the axis of x is a tangent at the point $x = -3$. There are curvilinear diameters corresponding to $2x = y^2 - 6$, and $y = \dfrac{2}{x}$. (Fig. 55.)

Ex. 16. To investigate all the curves expressed by the equation

$$y^3 - x^3 + ax^2 - b^2x = 0$$

The equation to the asymptote here, is $y = x - \dfrac{a}{3}$, when $a = 0$ the asymptote passes through the origin.

There are only two infinite branches, which when b^2 is greater or less than $\dfrac{1}{3}a^2$, will be found cutting the asymptote, and therefore extending on each side of it. When $b^2 = \dfrac{1}{3}a^2$ the ordinates parallel to the asymptote will have a diameter, and in this case the branches lie wholly on one side of it, for when $3b^2 = a^2$, the distance determining the intersection of the curve with its asymptote is infinite.

There will be a cusp in two cases, when $b = 0$, and when $a = \dfrac{1}{2}b$, in the latter case the cusp will be on the axis of x, at a distance $=\frac{1}{2}a$ from the origin. In the former case the cusp is at the origin.

There will exist a general centre in two cases, 1st, when $a = 0$, when the origin itself is a general centre, 2dly, when $9b^2 = 2a^2$, in which case the general centre is on the axis at a distance $=\frac{a}{3}$ from the origin.

When $b > 0$, there is a point of inflexion at the origin.

Since the proposed equation has two imaginary roots, except when the three roots are equal, the curve does not admit of oval or conjugate point. There are therefore only species 35, 37, 38, 42, or 45 of the enumeration to which it can be referred.

Ex. 17. Trace the curve

$$y^2 - xy - x^2y + x^3 + a^2 = 0$$

The equation referred to the analytical parallelogram gives two effective equations for x infinite, viz.,

$$x^3 - x^2y = 0, \text{ and } y^2 - x^2y = 0$$

whence $x = y$, and $y = x^2$, indicating hyperbolic branches approaching an asymptote which cuts the axis at an angle of $45°$, and also parabolic branches approaching a parabolic asymptote, of which the axis of y is a diameter.

Expressing y in a series, the terms are

$$y = x + \frac{a^2}{x^2} \&c.,$$

therefore the curve lies above the asymptote;

and since when $x = 0$, $y = \sqrt{-a^2}$, showing that the positive and negative hyperbolic branches are not joined together, it follows that the parabolic branches join the hyperbolic branches.

In this example if $a^2 = 0$, the equation becomes

$$y^2 - xy - x^2y + x^3 = 0 = (y - x)(y - x^2)$$

and the locus is a combination of the straight line and parabola.

The curve is the 53d of the enumeration.

If the term $-x^2y$ is also deficient in the equation, the curve is a divergent parabola whose axis cuts the axis of x at an angle of $22\frac{1}{2}°$, the branches lying on the negative side of the origin.

If the terms y^2 and $-x^2y$ are deficient, the curve is a trident.

Examples of Propositions producing Loci of the Third Order.

Ex. 1. At equal distances from the centre of a circle, ordinates to AB, the diameter, as NR, MP, are drawn. Then Q, the intersection of AP with NR, is the locus of a line of the third order.

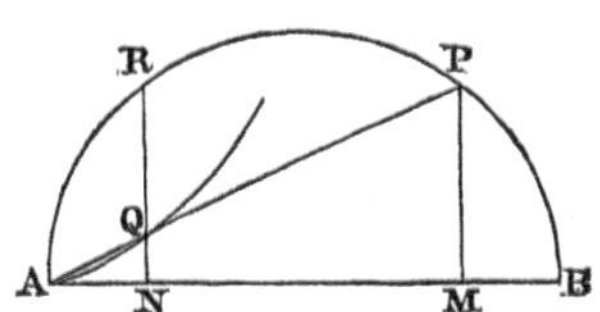

By similar triangles,

$$AN : NQ :: AM\,(= NB) : MP$$

$$\therefore MP^2 = \frac{NQ^2 . NB^2}{AN^2} = NB . AN$$

multiplying by $\dfrac{AN^2}{NB}$,

$$NQ^2 . NB = AN^3$$

or (if $AN = x$, $NQ = y$, $AB = a$),

$$(a - x)\,y^2 = x^3$$

the locus of a cissoid, whose asymptote passes through B, and whose vertex is at A.

Ex. 2. From one end of the diameter of a circle, straight lines as BA are drawn to meet the tangent at the other end. In

BA take AH always equal to the corresponding chord BI.   To find the locus of H,

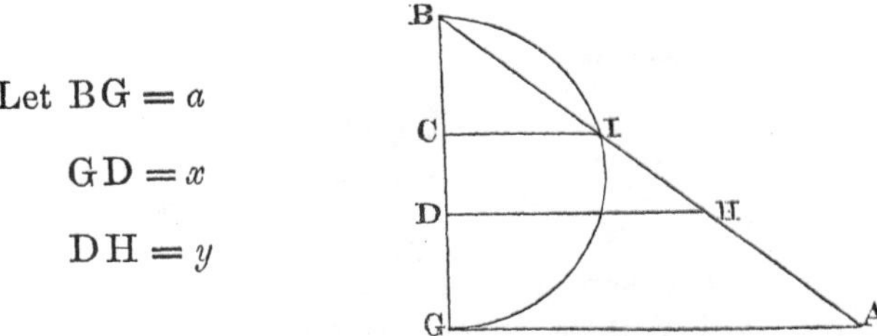

Let $BG = a$

$\qquad GD = x$

$\qquad DH = y$

Since $BI = AH$, drawing IC, HD parallel to the tangent, we have $BC = DG$,

and $$BC : CI :: BD : DH$$

or $$x : \sqrt{ax - x^2} :: a - x : y$$

whence $$xy^2 = (a - x)^3$$

which represents a defective hyperbola having a diameter, and the three equal roots show that the locus is a cissoid proper.

Ex. 3.   Q, A, N, are three flagstaffs standing on a plane. A person at a distance views them apparently at equal distances apart, although they are in reality at unequal distances.   What must be his path on the supposition that he moves towards the middle flagstaff, always preserving equal apparent angular distances between it and the other two?

If Q, A, N, the three flagstaffs be in a straight line, draw PM perpendicular from P a point in the path.   Join PQ, PA, PN, the angle QPN is bisected by PA.   Let

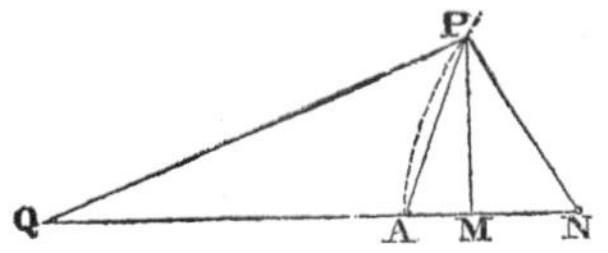

$$QA = a, \ AN = b, \ AM = x, \ MP = y$$

(by Euclid, VI. B.) $QP . PN = QA . AN + AP^2 = x^2 + ab + y^2$ (1)

also, $\qquad QP^2 = y^2 + (a + x)^2, \ \text{and} \ PN^2 = y^2 + (b - x)^2$

$\therefore$ from (1)

$$QP^2 . PN^2 = (x^2 + ab + y^2)^2 = \{y^2 + (a+x)^2\}\,\{y^2 + (b-x)^2\} \quad (2)$$

which equation reduced is

$$xy^2 + \frac{a-b}{2}\,y^2 = -x^3 - \frac{a^2 - 6ab + b^2}{2(a-b)}\,x^2 + abx \qquad (3)$$

In the particular case supposed, viz., that the flagstaffs stand in a line with one another, the equation (3) represents a combination of a circle and straight line, but in any other case the locus will be a defective hyperbola passing through the origin, and having a loop passing through Q or N.

Ex. 4.   Two straight lines, A B, H D, intersect at right angles in C. The radius B I revolving on B cuts C D in H. C Q perpendicular to B I being drawn, to find the locus of the point which is determined by taking on B I, H P always equal to Q B.

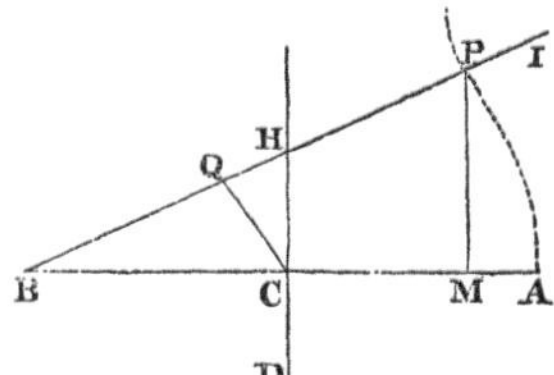

Let $BC = CA = a$.

Then B being the pole, the polar equation is evidently

$$r = a \sec \theta + a \cos \theta$$

To find the equation referred to rectangular axes, let $AM = x$, $MP = y$, then by similar triangles,

$$BH = \frac{BC . BP}{BM} = \frac{a\sqrt{(2a-x)^2 + y^2}}{2a - x}$$

also, $$HP = \frac{CM . BP}{BM} = \frac{(a-x)\sqrt{(2a-x)^2 + y^2}}{2a - x}$$

but $$BH . HP = BC^2$$

$$\therefore (a^2 - ax)\,\{(2a-x)^2 + y^2\} = a^2(2a-x)^2$$

whence $(a-x)y^2 = x^3 - 4ax^2 + 4a^2x$

The equation is that of a conchoidal hyperbola, whose asymptote passes through C. There is a point of inflexion when $x = \frac{2}{5}a$, and a conjugate point at B, therefore the species is 44, fig. 53.

Ex. 5. A point D is taken on the diameter AB of a given circle, and any perpendicular NN′ to AB being drawn, AN. AN′ are joined. Through D, parallel to AN, AN′, are drawn DP, DP′. Required the locus of P.

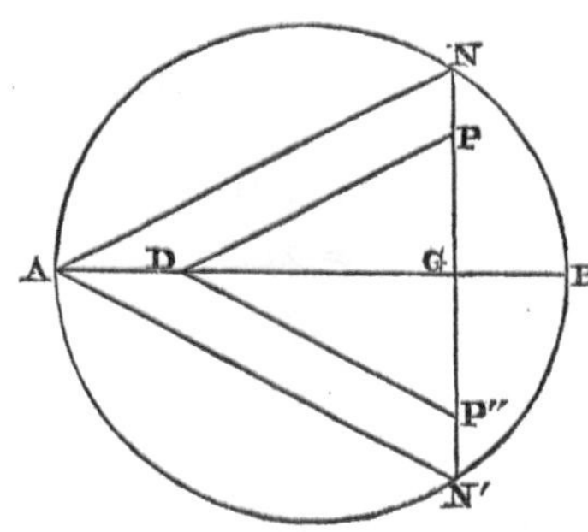

Let $AG = x$ $AB = a$

    $GP = y$ $AD = b$

Since $AG^2 : GN^2 :: DG^2 : GP^2$,

$\therefore x^2 : ax - x^2 :: (x-b)^2 : y^2$

or $\quad xy^2 + x^3 - (2b - a)x^2 + (b^2 + 2ab)x - ab^2 = 0$

The curve passes through B and twice through D, and its asymptote is the axis of y. Species 41.

Cramer *Analyse*, p. 441.

Ex. 6. The side BC of a right-angled triangle being produced, a radius revolves on A as a pole. Required the locus of a point in the revolving line whose distance from the point of intersection, D, is equal to the distance of D from C the vertex of the triangle.

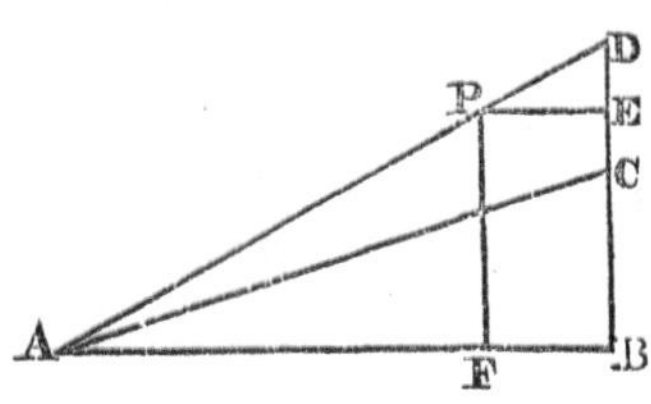

$AB = a$

$BC = b$

$AF = x$

$FP = y$

Draw FP perpendicular to AB

PE   to BD

By sim. tri.

$$BD = \frac{ay}{x}, \therefore CD = \frac{ay}{x} - b$$

But $$PD^2 = CD^2$$

$$\therefore \left(a - x\right)^2 + \left(\frac{ay}{x} - y\right)^2 = \left(\frac{ay - bx}{x}\right)^2$$

which equation reduced gives as that of the locus,

$$xy^2 - 2ay^2 + 2aby = -x^3 + 2ax^2 + (b^2 - a^2)\,x$$

from which it appears that the curve is a serpentine hyperbola without a diameter. When $x = a$, $y = b$, therefore the curve passes through C, which is a double point, and there is a loop between A and C. There is an asymptote when $x = 2a$.

The polar equation to this curve is obviously

$$r = a \sec \theta \mp a \tan \theta \pm b$$

Ex. 7. The vertex of a right angle slides on a straight line, AB, while one of its legs passes through a fixed point, H; to show that the intersection P of a perpendicular from a second fixed point, K, with the other leg, will be the locus of a line of the third order.

Let HNQ be a position of the right angle; draw HA per-

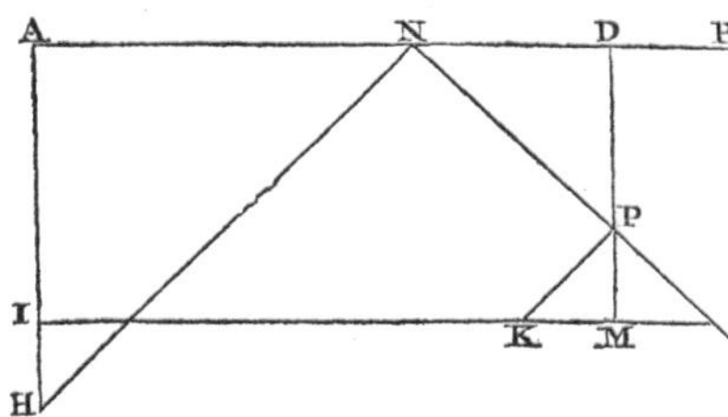

pendicular on AB, KP perpendicular on NQ, IKM parallel to AB, MPD perpendicular to IKM.

Let KM$=x$, MP$=y$, KI$=a$, AI$=b$, HA$=c$ $\therefore$ DP$=b-y$.

By sim . tri. KMP, NDP, HAN, NA$=\dfrac{cx}{y}$, ND$=\dfrac{by-y^2}{x}$

$$\therefore NA + ND = (a+x) = \frac{cx}{y} + \frac{by-y^2}{x}$$

Hence $$b y^2 - y^3 = x^2 y + a x y - c x^2$$

This equation, applied to the analytical triangle, gives

(1) $c x^2 - x^2 y = 0$, whence $y = c$, the equation to the asymptote.

(2) $- x^2 y - y^3 = 0$, two impossible asymptotes.

Hence the curve is a defective hyperbola; its species will depend on the relative position of AB, H, and K. If the line HK is drawn, and a semicircle described to the diameter HK, then, if AB intersects the semicircle twice, the curve is No. 34, a nodate hyperbola. If AB touches the semicircle, the locus is a cusped hyperbola (No. 35); if AB is outside the semicircle, the curve is a punctate hyperbola. Species 36.

Maclaurin, *Geom. Org.* p. 36, cor. 7.

Ex. 8. The corner of a page being doubled down, so that the sum of the edges is constant, to show that the locus of the angular point is a line of the third order.

Take the edges for axes, and their intersection as the origin. Let KPD be any position of the doubled part. Draw OCP, KCD, and let fall CB perpendicular to OM, CI to OK.

$$OM = x, \quad OB = \tfrac{1}{2} x$$
$$MP = y, \quad BC = \tfrac{1}{2} y$$

Since OCD, OCK, are right-angled triangles,

$$BC^2 = OB \cdot BD,$$
$$CI^2 = OI \cdot IK.$$

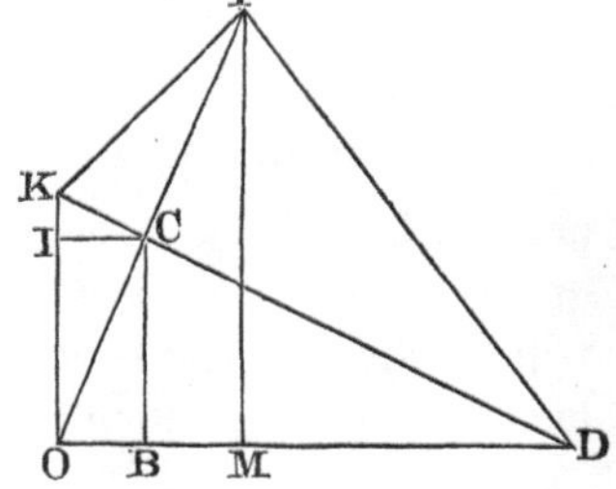

$$\therefore OD = (OB + BD) = \frac{x^2 + y^2}{2\,x}, \text{ and } OK = (OI + IK) = \frac{x^2 + y^2}{2\,y}$$

but $OD + OK = \text{constant} = a$,

$$\therefore x^3 + x^2 y + x y^2 + y^3 = 2\,a x y$$

The locus represents the loop of a defective hyperbola, having a double point at O, and the axes touch the curve at O. Referred to polar co-ordinates, if $OP = r$, the equation is

$$r = \frac{2a}{\sec\theta + \operatorname{cosec}\theta}$$

If $OD \cdot OK$ is constant, the locus represents a loop of a line of the fourth order, viz., a lemniscate.

If, instead of the sides, it is the crease which is constant, the locus of the point P will be the loop of a line of the sixth order; for if $r = OC$, $COD = \theta$

Since $KC + CD = \text{constant} = a$,

$$r\cot\theta + r\tan\theta = a$$

$$\therefore\ r = \frac{a}{\cot\theta + \tan\theta}$$

or in rectangular co-ordinates $(x^2 + y^2)^3 = a^2 x^2 y^2$

This is one of the curves called " the roses" by the Abbate Guido Grandi, in his book called *Flores Geometrici*, where its equation is given in the form

$$r = \sin 2\theta$$

Ex. 9. Tangents are drawn to a circle, cutting the diameter produced (as IB). From a point H, which divides the tangent in a given ratio $m : n$, a perpendicular to the tangent is drawn, which cuts the ordinate or ordinate produced, in P. To find the locus of the intersection P.

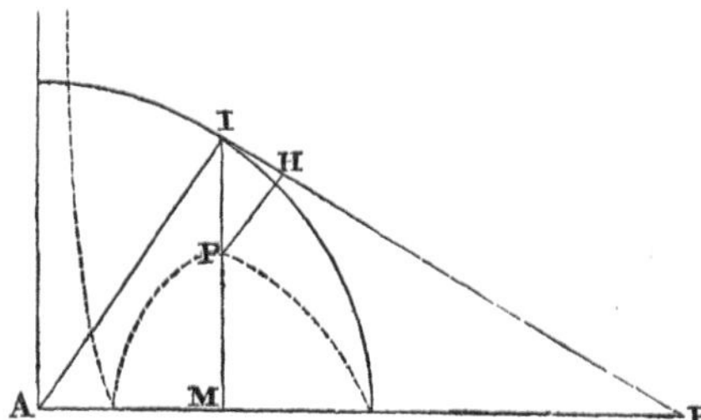

Let A be the centre of the circle.

Let $AM = x$, $MP = y$,
$AI = a$.

Then, since $AM : MI :: AI : IB$, $x : \sqrt{a^2 - x^2} :: a : \dfrac{a}{x}\sqrt{a^2 - x^2}$

also $BH : HI :: m : n$, $\therefore IB : HI :: m + n : n$

$$\therefore HI = \frac{na\sqrt{a^2 - x^2}}{(m+n)\,x}$$

and by sim. tri. $AM : AI :: HI : IP = \dfrac{na^2\sqrt{a^2 - x^2}}{(m+n)\,x^2}$

$$\therefore MP = y = \sqrt{a^2 - x^2} - \frac{na^2\sqrt{a^2 - x^2}}{(m+n)\,x^2}$$

If $x = a$, or $a\sqrt{\dfrac{n}{m+n}}$, in either case $y = 0$; and if $x = 0$,

$y = \pm\infty$.

If $x \gtrless a$, y is imaginary. The curve is therefore a defective hyperbola whose diameter coincides with AB, and whose asymptote passes through A.

Ex. 10. The base and altitude of a triangle being given, to find the locus of the intersection of the bisectors of the base angles.

Let ABC represent a position of the triangle; let AP, BP, bisecting the angles A, B, meet in P. Take the middle point of the base, O, as origin.

Let $OQ = x$

$\qquad OR = x'$

$\qquad PQ = y = (a - x)\tan PBA = (a + x)\tan PAB$

$\qquad CR = y' = (a - x')\tan CBA = (a + x')\tan CAB$

$\qquad AO = OB = a.$

Since by hypothesis $CAB = 2\,PAB$, and $CBA = 2\,PBA$,

$$\tan CAB = \frac{2\tan PAB}{1 - \tan^2 PAB}, \text{ and } \tan CBA = \frac{2\tan PBA}{1 - \tan^2 PBA}$$

from whence, by substituting these values,

$$\frac{y'}{a+x'} = \frac{2y\,(a+x)}{(a+x)^2 - y^2}, \text{ and } \frac{y'}{a-x'} = \frac{2y\,(a-x)}{(a-x)^2 - y^2}$$

eliminating x' and making y', which is the constant altitude of the triangle $= 1$, we have

$$2\,x^2 y - y^2 - x^2 - 2\,a^2 y + a^2 = 0$$

Here, when $x = \pm\, a$, $y = 0$, whence the curve passes through A and B; and if $x = 0$,

$$y = -\,a^2 \pm \sqrt{a^4 + a^2}$$

showing that the curve exists below as well as above the axis of x.

Solving the equation for y,

$$y = x^2 - a^2 \pm \sqrt{x^4 - (2\,a^2 + 1)\,x^2 + a^4 + a^2}$$

which shows a parabolic diameter bisecting chords parallel to the axis of y, and since the radical is always possible, if x is less than a, it follows that the curve is an oval, belonging to a parabolic hyperbola.

The portions of the curve outside the triangle are the loci of the intersection of the bisectors of the external angles of the triangle made by producing its sides, since the expressions already used for the co-ordinates of intersection still hold good.

To find the infinite branches, let the equation be referred to the analytical parallelogram; on the hypothesis of x infinite, we have

$$2\,x^2 y = x^2, \quad \text{whence} \quad y = \tfrac{1}{2} \qquad (1)$$

$$2\,x^2 y = y^2, \quad \text{whence} \quad 2\,x^2 = y \qquad (2)$$

From (1) it appears there is a rectilinear asymptote parallel to the axis of x, at a distance $= \tfrac{1}{2}\,y'$.

From (2) there is an asymptotic parabola, the axis of which is the axis of y.

Solving the equation for x, we have

$$x = \pm \sqrt{\frac{y^2 + 2a^2 y - a^2}{2y - 1}}$$

which being always possible for large positive values of y, shows the existence of branches of the curve accompanying the asymptote to infinity.

The curve here considered is not contained in Newton's enumeration nor in Stirling's commentary, but is noticed in Nicole's paper on the projections of the divergent parabola with an oval, in the *Mém. de l'Académie* for the year 1731. See fig. at p. 63.

Ex. 11. The ordinate to the diameter of a circle is produced to a point P, so that the rectangle under the absciss and such produced ordinate, shall equal the rectangle under the diameter of the circle and its ordinate. To show that the locus of P is a line of the third order,

Let $AM = x$

$MQ = y'$

$AB = a$

$MP = y$

By hypothesis, $xy = ay'$

and $y' = \pm \sqrt{ax - x^2}$

hence $y^2 x + a^2 x - a^3 = 0$

The curve is Species 62, fig. 71.

Ex. 12. Required the locus of intersections of the tangent to a parabola with the perpendicular on it from the origin at the vertex.

The equation to the tangent of the parabola being

$$y' = \frac{a^{\frac{1}{2}}}{x^{\frac{1}{2}}} (x' + x) \qquad (1)$$

that of a perpendicular on it is

$$y_{\prime} = -\frac{x^{\frac{3}{2}}}{a^{\frac{1}{2}}}\, x_{\prime} \qquad (2)$$

where $x_{\prime}$ and $y_{\prime}$ are the co-ordinates of the perpendicular on the tangent, and at their common intersection we have $x' = x_{\prime}$, and $y' = y_{\prime}$; therefore

$$y_{\prime} = \frac{a^{\frac{1}{2}}}{x^{\frac{1}{2}}}\, (x_{\prime} + x)$$

And since from (2),

$$\frac{x^{\frac{1}{2}}}{a^{\frac{1}{2}}} = -\frac{y_{\prime}}{x_{\prime}} \quad \text{we find} \quad x = \frac{a\,y_{\prime}^{2}}{x_{\prime}^{2}}.$$

whence equation (1) becomes

$$y_{\prime} = -\frac{x_{\prime}}{y_{\prime}}\left(x_{\prime} + \frac{a\,y_{\prime}^{2}}{x_{\prime}^{2}}\right) \text{ or } y_{\prime}^{2} = -\frac{x_{\prime}^{3}}{a + x_{\prime}}$$

representing a cissoid whose cusp is at the origin, and whose asymptote is the line $x = -a$.

If the origin be taken on some other point in the parabola than the vertex, the curve will be the 35th species, fig. 45.

Ex. 13. From any point K of a right line AK given in position, straight lines KD, KC, are drawn to two fixed points C,D. In KD take KP, KP′, each equal to KC. The points P, P′, will trace out a line of the third order.

Draw DB $(= a)$,
  CA $(= b)$
perpendicular to AK.

 Let AB $= c$,
DM$= x$, MP $= y$.

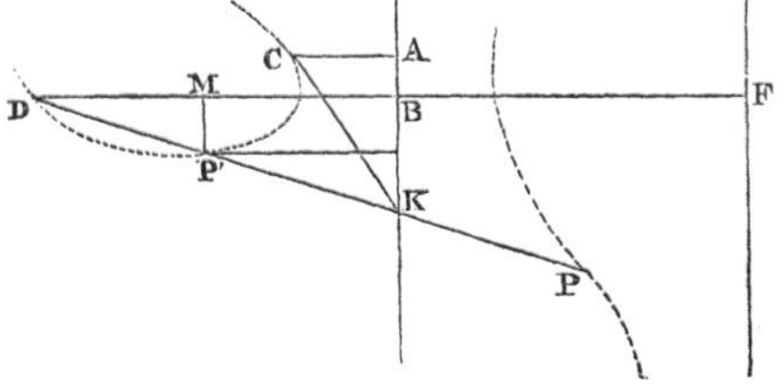

From the similar triangles DMP, DBK,

$$BK = \frac{a\,y}{x}, \quad AK = c + \frac{a\,y}{x}, \quad \text{and } CK^{2} = \left(c + \frac{a\,y}{x}\right)^{2} + b^{2}$$

Also $\quad \mathrm{P}\,\mathrm{K}^2 = \left(\dfrac{ay}{x} - y\right)^2 + (a - x)^2$, and $\mathrm{C}\,\mathrm{K}^2 = \mathrm{P}\,\mathrm{K}^2$

$$\therefore \quad x^3 - 2\,a\,x^2 - (b^2 + c^2 - a^2)\,x = 2\,a\,y^2 - x\,y^2 + 2\,a\,c\,y$$

which equation belongs to a defective hyperbola without a diameter, having a conjugate oval passing through C and D.

Equating the cubic terms to 0, we have

$$2\,a\,y^2 - x\,y^2 = 0, \quad \text{or} \quad 2\,a = x$$

which gives the rectilinear asymptote, passing through F, if $\mathrm{BF} = a$.

It is a property of this curve that the straight line A K bisects all lines terminated by the curve, provided they all meet in the point D. If the fixed points do not lie on the same side of the given straight line, the infinite branch will pass through one of the fixed points, and the oval through the other.

The curve is species 33 of the enumeration.

Ex. 14. If from any three points, lines be drawn to intersect in a fourth point, so that one of them shall be always an arithmetic mean to the other two, to find the locus of the point of intersection.

The locus required is the 29th, 33d, 34th, 40th, 41st, or 44th species of the enumeration.

On the same hypothesis to find the locus, when the lines are drawn so that the rectangle under two of them shall equal the square of the third.

The locus required is the 33d, 34th, 37th, 40th, or 45th species, according as the relative position of the three points varies.

Leybourn's *Mathematical Repository*, vol. iii. p. 227.

Ex. 15. A straight line parallel to the diagonal of a rhombus is drawn through one of its angular points. Required the locus of centres of a system of conic sections, each of which touches the line parallel to the diagonal, and passes through the remaining angular points of the rhombus.

The required locus is a redundant hyperbola whose asymptotes form a triangle, with a cusp at the centre of the rhombus. Species 12, fig. 24.

Hearn's *Researches on Curves of the Second Order*, p. 45.

Ex. 16.   From a given point S in the diameter AB produced of a given circle, let any line be drawn cutting the circle in N and D, draw AD, BN intersecting in H.   If in AD produced there is always taken DP = DH, to determine the curve traced by P.

From S draw the tangent ST, then the perpendicular TG is given in position, and is the locus of H.

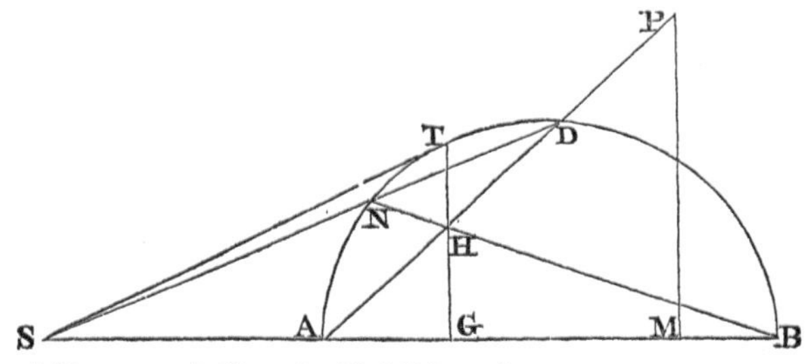

Let $AB = a$, $AG = b$, $PAM = \theta$.

Since $AP = AD + DP = AD + DH = 2\,AD - AH$

we have if $AP = \varrho$, $AD = r$

$$\varrho = 2r - b \sec \theta \quad . \quad . \quad . \quad (1)$$

but by the equation to the circle, $r = a \cos \theta$

Hence (1) becomes

$$\varrho = 2a \cos \theta - b \sec \theta$$

in rectangular co-ordinates, if $AM = x$, $MP = y$

$$(b + x)\, y^2 = (2a - b)\, x^2 - x^3$$

(Species 41.)

Ex. 17.   To determine the species of the curve traced by the extremity of a portion of the linear sine of an angle θ, which is taken always equal to the linear tangent of half that angle.

By the hypothesis, $y = \tan \dfrac{\theta}{2}$

$$= \pm \sqrt{\dfrac{1 - \cos \theta}{1 + \cos \theta}}$$

$$= \pm \sqrt{\dfrac{x}{2 - x}}$$

which equation represents species 63 of the enumeration.

Miscellaneous Problems having reference to Lines of the Third Order.

Prob. 1.   To describe lines of the third order by continuous motion.

Take a grooved ruler of sufficient length and let it revolve on the fixed pin A, while one end of it slides along a straight line BC.   Let one end of a string of equal length with the ruler be fastened at the point R, pass over the sliding pin P, and being stretched along the ruler's edge, have its other end attached to the ruler's farther end.   Then the point P will describe a line of the third order, whose species will depend on the position of the point R.

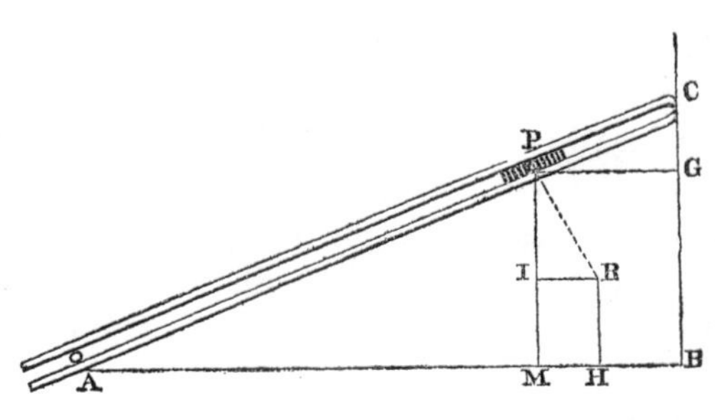

Draw the perpendiculars A B, P G, P M, R H, R I.

Let $AM = x$,
$\quad MP = y$,
$\quad AH = a$,
$\quad HR = b$,
$\quad AB = c$

By sim. tri. APM, PCG,

$$AM : AP :: PG : PC = PR = \sqrt{(y-b)^2 + (a-x)^2}$$

$$x : \sqrt{x^2 + y^2} :: c - x : \frac{(c-x)\sqrt{x^2+y^2}}{x}$$

$$\therefore \quad (c-x)^2 (x^2 + y^2) = x^2 \{(y-b)^2 + (a-x)^2\}$$

$$\text{and} \qquad xy^2 - \frac{c}{2}y^2 - \frac{b}{c}x^2 y = \frac{a-c}{c}x^3 + \frac{c^2-a^2-b^2}{2c}x^2$$

is the general equation to all lines capable of description by this method. First let us suppose that R is a point in AB, in this case $b = 0$; if R coincides with B, $a = c$, and the curve degenerates into straight lines. But when R is between A and B, the curve is a conchoidal hyperbola with a conjugate point at A, Species 44. When R falls on the left of A, so long as its distance from A is less than AB, the curve is still a conchoidal hyperbola, but with its convex side towards the conjugate point, Species 43. When this distance is exactly AB, the curve is a cissoid; if the distance exceed AB, the curve is Species 41. Next, supposing R to fall on the right of B, at a less distance than AB; the curve is now a cruciform hyperbola, Species 18. When R is exactly at the distance AB, the curve is the 30th species; when at a greater distance, the 19th species. The asymptote in all these cases is the same, viz. the line bisecting AB and parallel to BC.

But if R lies in the line BC, the locus degenerates into a conic hyperbola with a straight line.

Suppose a conic parabola described with AB as axis, B as the vertex, and A as the focus; and suppose also a circle on AB as diameter, and that R falls between the parabola and BC. The curve will be a redundant hyperbola; when R is on the parabola the curve is a parabolic hyperbola; when within it, a defective hyperbola, which becomes the 34th, 35th, or 36th species, according as R is without, upon, or within the circumference of the circle: in all these cases the curve has no diameter.

But R may fall on the right of BC, in which case the curve

will be of the 7th, 8th, or 25th species, according as its distance from BC varies. In fact, by this method all the non-parabolic curves which have a double point may be drawn.

Prob. 2. BE represents the hyperbolic branch of a conchoidal curve, species 44, of the enumeration, A its conjugate point, C the intersection of the asymptote with the axis. If now BF the vertical tangent, be drawn, and on AB, as diameter, a circle be described, and a line AF be supposed to revolve on A, cutting the circle in D, the branch in E, and the tangent in F, to show that the ratio DE : EF is constant.

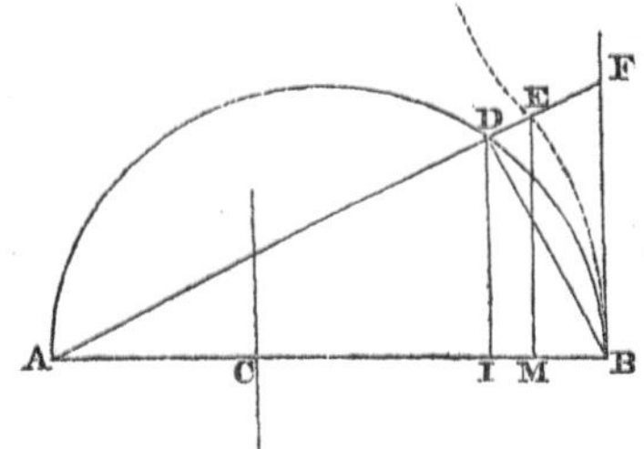

Join DB

Draw D I, E M, perpendicular to A B.

By the property of the curve

$$MC . ME^2 = MB . MA^2$$

$$\therefore \quad MC : MB :: (MA^2 : ME^2 :: AD^2 : DB^2) :: AI : IB$$

comp. $\quad AB : IB :: CB : MB$

alt. $\quad AB : CB :: IB : MB :: DF : EF$

div. $\quad AC : CB :: \qquad :: DE : EF$

And the ratio AC : CB being constant, so also is the ratio DE : EF, whatever the position of the line AF.

Prob. 3. If from the cusp of a cissoid, radii be drawn to intersect the branches, perpendiculars through the points of intersection will envelope a parabola.

The equation to the cissoid being $(a + x) y^2 = - x^3$

$$y = - \frac{x^2}{y} - \frac{a y}{x} \qquad (1)$$

Let the equation to any assumed radius be

$$m y = - x \qquad (2)$$

Then that of a perpendicular on it will be

$$y = m x + b \qquad (3)$$

Since at the intersection these three equations are simultaneously true for that point,

$$\frac{- x^2}{y} - \frac{a y}{x} = m x + b, \text{ and } m = - \frac{x}{y} \therefore b = \frac{a}{m}$$

Hence (3) becomes $y = m x + \dfrac{a}{m}$

which is an equation to a line touching the parabola

$$y^2 = 4 a x$$

the asymptote and directrix coincide, the cissoid lying in the negative and the parabola in the positive region.

Prob. 4. In the semicircle AGB, of which MG is an ordinate, the area AMP is always taken equal to the corresponding segment ANG. Required the curve generated.

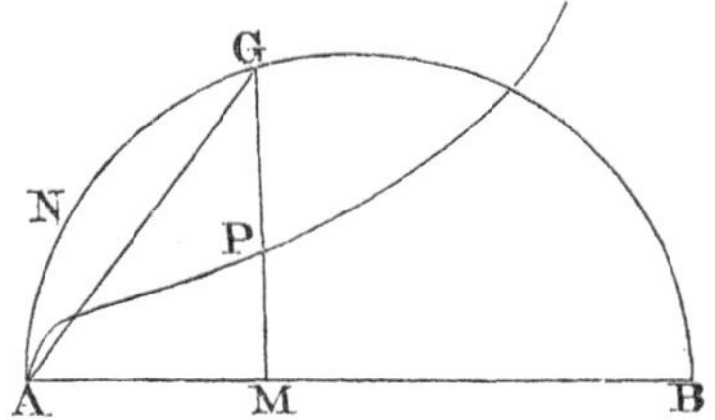

Let $AM = x$
$MP = y$
$MG = y'$
$rad = 1$

$$\text{Then } \int y\,dx = \text{ANG} = \text{ANGM} - \text{AGM}$$

$$= \int y'\,dx - \frac{y'x}{2}$$

$$= \int \sqrt{2x - x^2}\,dx - \frac{x\sqrt{2x - x^2}}{2}$$

$$\therefore y\,dx = \sqrt{2x - x^2}\,dx - \frac{\sqrt{2x - x^2}\,dx}{2} - \frac{x(1 - x)\,dx}{2\sqrt{2x - x^2}}$$

Hence $y = \dfrac{x}{2\sqrt{2x - x^2}}$ is the equation to the curve, which represents a hyperbolism of the ellipse, whose asymptote passes through B, and whose vertex is A. It is evident that the area of this curve between the vertex and asymptote is equal to the circle described on AB as diameter.

Prob. 5. Required the curve in which the area is always equal to the fourth power of the ordinate.

$$\text{We have} \qquad y^4 = \int y\,dx$$

$$4y^3\,dy = y\,dx$$

$$4y^2\,dy = dx, \text{ and integrating}$$

$$y^3 = \frac{3}{4}x + C$$

which is a case of the cubic parabola.

Prob. 6. Required the curve in which the area is always equal to twice the rectangle under the co-ordinates?

$$\text{Since} \qquad \int y\,dx = 2xy$$

$$y\,dx + 2x\,dy = 0$$

$$2xy\,dy + y^2\,dx = 0, \text{ and integrating}$$

$$2xy^2 = C$$

the curve is a hyperbolism of the parabola, the origin of co-ordinates being taken at the intersection of the asymptotes.

Prob. 7. Required the curve in which the area is always equal to three-fifths the rectangle under the co-ordinates.

Since
$$\int y\,dx = \frac{3}{5}\,xy$$

$$y\,dx = \frac{3}{5}\,y\,dx + \frac{3}{5}\,x\,dy$$

$$\therefore \quad \frac{2}{5}\frac{dx}{x} = \frac{3}{5}\frac{dy}{y}$$

$$\therefore \quad \frac{2}{5}\log x = \frac{3}{5}\log y$$

$$\therefore \quad x^2 \; \propto \; y^3$$

the curve is a semicubic parabola.

Prob. 8. Required the curve in which the area is always equal to three-fourths the rectangle under the co-ordinates.

We have here
$$\frac{1}{4}\log x = \frac{3}{4}\log y$$

$$x \; \propto \; y^3$$

the curve is a cubic parabola.

Prob. 9. To find the area of the hyperbolism of the hyperbola included between the curve and parallel asymptotes.

The equation being
$$(a^2 - x^2)\,y = a^3$$

where $2a$ is the distance between the parallel asymptotes,

$$\int y\,dx = \int \frac{a^3\,dx}{a^2 - x^2} = \frac{a^2}{2}\log\frac{a+x}{a-x} = \text{the area required.}$$

Hence the whole area, when $x = a$, is infinite.

Prob. 10. To find the area of the hyperbolism of the ellipse taken between the curve and its asymptote.

The axis of the ellipse being taken as the parameter of the curve, the required area is equal to four times the generating ellipse.

Prob. 11. To find the area of the cissoid between the curve and its asymptote.

The required area is equal to three times the generating circle.

Prob. 12. To find the area of the loop of the folium of Des Cartes.

The equation being

$$x^3 - 3axy + y^3 = 0$$

the required area will be $= \dfrac{3}{2} a^2.$

Prob. 13. To find the point of greatest curvature in the semicubic parabola,

$$ay^2 = x^3$$

the radius of curvature being $= - \dfrac{4 a^{\frac{1}{2}} x^{\frac{1}{2}}}{3} \left(1 + \dfrac{9x}{4a} \right)^{\frac{3}{2}}$

which is 0 when $x = 0$; the required point is the vertex, where the curvature is infinite.

Prob. 14. To find the point of greatest curvature in the cubic parabola

$$x^3 = a^2 y$$

the required point occurs when $x : y : : 3 \sqrt{5} : 1$

In the cubic hyperbola

$$xy^2 = a^3$$

the point of greatest curvature occurs when

$$x : y :: \pm \sqrt{5} : 1$$

Newton, Anal. Geom. p. 457.

Prob. 15.   To find the points of inflexion in the curve

$$xy^2 + ax^2 + b^3 = 0$$

When a and b are of the same sign, the curve is a parabolic hyperbola, species 53 (fig. 61).

There are points of inflexion when

$$x = \mp b \left\{ (3 + \sqrt{12}) \frac{b}{a} \right\}^{\frac{1}{2}}, \quad y = \pm \left\{ \left(4 + \frac{8}{\sqrt{3}} \right) ab^3 \right\}^{\frac{1}{4}}$$

When a and b are of contrary signs, the curve is species 56 (fig. 64); and there will be points of inflexion when

$$x = \mp b \left\{ (3 - \sqrt{12}) \frac{b}{a} \right\}^{\frac{1}{2}}, \quad y = \pm \left\{ \left(4 - \frac{8}{\sqrt{3}} \right) ab^3 \right\}^{\frac{1}{4}}$$

Prob. 16.   To find the point of inflexion in the curve

$$x^2 y + bxy - ax^2 + aby - a^2 x = 0$$

the tangential equation is here $aby - a^2 x = 0$; and since this will divide the row of terms superior to it (p. 97), which is, $bxy - ax^2 = 0$, there is a point of inflexion at the origin.

Prob. 17.   A perpendicular to the axis of a curve is drawn through a fixed point in the axis, and intersects the tangent to any point P, in the point Q.   The length of this perpendicular and the intercept PQ together is double the length of the curve between the vertex and P.   Required the equation and species of the curve.

If the distance from the fixed point to the vertex is called a, the equation is found to be

$$y = \pm \frac{1}{3\sqrt{a}} (4a - x) \sqrt{x - a}$$

which equation represents a nodate parabola, species 68 of the enumeration.

Leybourn's Mathematical Repository, New Series, vol. iv., p. 100.

Prob. 18.   A straight line being supposed to revolve about a fixed point P in the absciss of a line of the third order, so as to cut the curve in three points, and tangents to these points being drawn to meet the absciss, to show that the sum of the reciprocals of the subtangents is constant.

Let the equation be arranged in the form

$$y^3 - (ax + b) y^2 + (cx^2 + dx + e) y \ \&c. = 0$$

Let y, y', y'', be the three values of ordinates corresponding to a given absciss, then (p. 38)

$$yy'y'' = fx^3 - gx^2 + hx - k \dots \dots (1)$$

and differentiating

$$y'y''dy + yy''dy' + yy'dy'' = 3fx^2dx - 2gxdx + hdx \dots (2)$$

dividing (2) by (1) and also by dx,

$$\frac{dy}{y\,dx} + \frac{dy'}{y'\,dx} + \frac{dy''}{y''\,dx} = \frac{3fx^2 - 2gx + h}{fx^3 - gx^2 + hx - k}$$

which proves the proposition, because the fraction is constant, x being invariable while P is fixed, whatever the angle of ordination, and therefore the sum of the reciprocals of the subtangents is also constant.

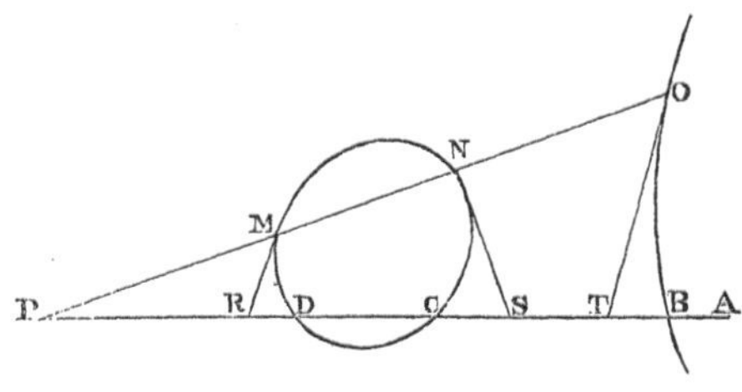

Thus if $AP = x$, PM, PN, PO, = the three values of y, y', y'', MR, NS, OT, the three tangents, the sum $\dfrac{1}{PR} + \dfrac{1}{PS} + \dfrac{1}{PT}$ is constant.

Prob. 19. The same things being supposed as in the last proposition, if the fixed line AP also cuts the curve in three points, D, C, B, the sum

$$\frac{1}{PR} + \frac{1}{PS} + \frac{1}{PT} = \frac{1}{PD} + \frac{1}{PC} + \frac{1}{PB}$$

that is, the sum of the reciprocals of the subtangents, will be equal to the sum of the reciprocals of the segments of the absciss between the fixed point P and the curve.

This is evident from the consideration that when PO coincides with PB, $PD = PR$, $PC = PS$, and $PB = PT$; therefore

$$\frac{1}{PD} + \frac{1}{PC} + \frac{1}{PB} = \frac{1}{PR} + \frac{1}{PS} + \frac{1}{PT}$$

which sum is constant at whatever angle the ordinates are inclined to the fixed axis. When the segments are on contrary sides of P, the terms will be affected by contrary signs.

Prob. 20. A straight line PA revolving about P cuts a line of the third order in three points, B, C, D; and in PA, Q is taken so that PQ is a harmonic mean between PB, PC, PD. To show that the locus of Q will be a straight line.

In the case of a right line PA, drawn from a fixed point P, which cuts any number of right lines given in position in the points E, F, G, H, &c., if in PA there is taken a point K such that $\dfrac{1}{PK}$ always $= \dfrac{1}{PE} \mp \dfrac{1}{PF} \mp \dfrac{1}{PG} \mp \dfrac{1}{PH}$ &c., then, as is well known, the locus of K will be a straight line.

Assuming PMNO (see the last diagram) to be fixed, and PA to revolve, then the tangents MR, NS, OT, are given in position; and if a point K is taken so that

$$\frac{1}{PD} + \frac{1}{PC} + \frac{1}{PB} \left(= \frac{1}{PR} + \frac{1}{PS} + \frac{1}{PT} \right) = \frac{1}{PK}$$

the locus of K will be a straight line; and if we take another point Q so that $PQ = 3PK$, we have

$$\frac{1}{PD} + \frac{1}{PC} + \frac{1}{PB} = \frac{3}{PQ}$$

∴ when PQ is a harmonic mean between the three lines PD, PC, PB, the locus of Q is a straight line.

Cotes on the Nature of Curves; Maclaurin, Lineæ Geometricæ, p. 460; *Salmon's Higher Plane Curves,* Art. 56.

Prob. 21. Through three points in a line of the third order in the same straight line tangents are drawn which also cut the curve. These points which they cut will also be in a straight line.

This and the six subsequent propositions are deductions from the three preceding propositions.

Prob. 22. If from a point in the curve a pair of tangents be drawn, and from either point of contact another pair of tangents be drawn, the line joining the last-named points of contact will pass through the remaining point of contact.

Maclaurin, Lin. Geom., Prop. 7, Cor. 1.

Prob. 23. If from a point of inflexion a pair of tangents be drawn, as AF, AG, and the line joining the points of contact be produced to meet the curve in H, AH touches the curve in H.

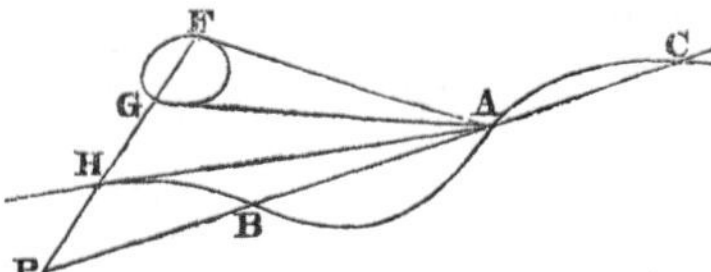

Not more than three tangents can be drawn from a point

of inflexion (besides that which touches and cuts it at the same point), and the three points of contact are in the same straight line.

Prob. 24. If from a point of inflexion A (see the last figure), three tangents are drawn, the straight line joining the points of contact will cut harmonically any straight line drawn through the point of inflexion and terminated by the curve.

For let F G H, the line joining points of contact, be produced to meet the line C A B drawn through the point of inflexion in P, then, since the three tangents all meet in one point,

$$\frac{1}{PA} + \frac{1}{PB} + \frac{1}{PC} = \frac{3}{PA}, \text{ or } \frac{1}{PB} + \frac{1}{PC} = \frac{2}{PA}$$

that is, P A is a harmonic mean between P B and P C.

Prob. 25. A straight line joining two points of inflexion either passes through a third point of inflexion or is parallel to the ultimate direction of the infinite branches. For draw P Q joining P and Q, two points of inflexion, and produce it to meet the curve in R, then, to show that R must be also a point of inflexion, let a tangent be drawn through R, cutting the curve in another point S, if possible; and by Prop. 21 P, Q, and S are in the same straight line with R, which is impossible in a line of the third order. Therefore R is a point of inflexion.

Hence we infer that the infinite branches of the divergent parabolas are ultimately parallel to the line joining their points of inflexion.

Prob. 26. A line drawn through a point of inflexion A, and parallel to the asymptote, meets the curve again in B. If a tangent at B is drawn cutting the curve in C, the line AC passes through the point where the curve and asymptote intersect.

Maclaurin, Lin. Geom., p. 10.

Prob. 27. If four tangents are drawn from P, a point in the curve, lines joining the points of contact will intersect in some point of the curve, and any straight line drawn from P will be cut harmonically by the curve and the lines joining two points of contact.

Maclaurin's Fluxions, art. 402.

Prob. 28. The equation to a line of the third order being given, to draw a chord parallel to one of the axes of co-ordinates, so as to be bisected by the curve.

The equation being arranged in the form

$$x^3 + (a + by)\,x^2 + \&c. = 0$$

or

$$y^3 + (a + bx)\,y^2 + \&c. = 0$$

according as the chord is required to be parallel to one or other of the axes, let the locus be constructed of the equation

$$3\,x = -a - by$$

or

$$3\,y = -a - bx$$

If the straight line thus drawn cuts the curve, it intersects it at a point which bisects the required chord. If it does not cut the curve, no such chord can be drawn. When the coefficient of x^2 or $y^2 = 0$, the required chord coincides with the axis.

Prob. 29. Given the three asymptotes of a redundant hyperbola intersecting so as to form a triangle; required the locus of the points of osculation within the triangle.

The condition for a point of osculation is (*Gregory's Examples*, ch. x. p. 166),

$$\left(\frac{d^2 u}{dx\,dy}\right)^2 - \left(\frac{d^2 u}{d\,x^2}\right)\left(\frac{d^2 u}{dy^2}\right) = 0$$

Referring the curve to two of its asymptotes as axes, their intersection being the origin, its equation takes the form

$$u = ax^2 y + 2\,bxy + cxy^2 = h$$

hence
$$\left(\frac{du}{dx}\right) = 2\,axy + 2\,by + cy^2$$

$$\left(\frac{du}{dy}\right) = 2\,cxy + 2\,bx + ax^2$$

$$\left(\frac{d^2u}{dx^2}\right) = 2\,ay, \quad \left(\frac{d^2u}{dy^2}\right) = 2\,cx, \quad \left(\frac{d^2u}{dx\,dy}\right) = 2\left(ax + b + cy\right)$$

Therefore for a point of osculation,

$$(ax + b + cy)^2 - acxy = 0$$

or
$$a^2x^2 + acxy + c^2y^2 + 2\,abx + 2\,bcy + b^2 = 0 \ \ldots \ (1)$$

This equation represents an ellipse, the position of which is determined by its touching the middle points of the three sides of the triangle. For it appears from (1) that it touches the axes at distances $-\dfrac{b}{a}$, and $-\dfrac{b}{c}$ from the origin, and the equations to the three asymptotes are

$$x = 0, \ \ y = 0, \ \ ax + 2b + cy = 0 \ \ \ldots \ \ (2)$$

giving $-\dfrac{2b}{a}$ and $-\dfrac{2b}{c}$ as the intercepts cut off by the third asymptote. If the values $-\dfrac{b}{a}$ and $-\dfrac{b}{c}$ are substituted for x and y in (1), the value of $\dfrac{dy}{dx}$ derived from the resulting equation coincides with the value derived from the equation to the third asymptote, viz. $-\dfrac{a}{c}$, and as these values of x and y satisfy both the equation to the ellipse and that to the asymptote, the ellipse touches the third side of the triangle in its middle point.

Prob. 30. To investigate the properties of the logocyclic curve.

The name logocyclic has been given by Dr. Booth to a peculiar combination of Des Cartes' folium with a common parabola. The equation to the former being for rectangular co-ordinates,

$$y^2 (2a - x) = x (a - x)^2$$

F being the origin, and FO being taken $= a$, the course of the curve will be as shown in the diagram, where QO represents the branch of the parabola whose focus is at F, and whose vertex is at O.

If F be made the pole and FO the prime radius, r and θ being the polar co-ordinates, since $\frac{y}{x} = \tan \theta$, $\frac{r}{x} = \sec \theta$, $r = \sqrt{x^2 + y^2}$, the polar equation is

$$r = a (\sec \theta \pm \tan \theta)$$

Thus, if the radius vector cuts the curve in the points R, R',

$$\mathrm{F R} = a (\sec \theta + \tan \theta), \ \mathrm{F R'} = a (\sec \theta - \tan \theta),$$

and the product $\mathrm{F R} \cdot \mathrm{F R'} = a^2 (\sec^2 \theta - \tan^2 \theta)$, is constant, and $= a^2$.

If through R and R' normals RQ, R'Q be drawn to the curve intersecting one another, they are equal.

If tangents be drawn at R, R' meeting in V, they also are equal, and equally inclined to the chord RR'. These three properties of this curve are also properties of the circle.

The locus of the points Q is a parabola whose focus is F, whose vertical focal distance is a, and whose directrix is the asymptote of the curve. By means of this parabola and this its allied curve, Dr. Booth has, in a recent communication to the proceedings of the Royal

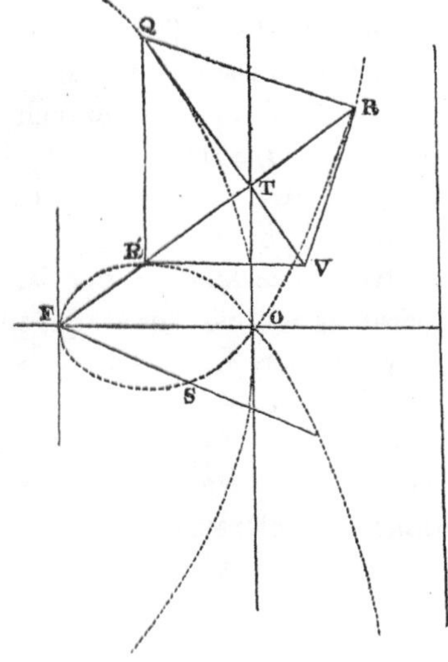

Society, exhibited the whole theory of logarithms in a geometrical form singularly comprehensive and beautiful. From the analogy of its properties with those of the circle, and from its use in illustrating the logarithmic theory, he has called it the logocyclic curve.

To represent numbers and their logarithms by the logocyclic curve and its conjugate parabola, Dr. Booth gives the following rule : — A parabola whose focal vertical distance is $a = 1$ being drawn, and also its logocyclic curve, let a radius vector be drawn to the latter equal to the number n. Then $n = \sec \theta + \tan \theta$.

Let this line meet the vertical tangent in T, the parabolic arc $OQ-QT$ is the logarithm of n.

It is clear that the infinite branch of the curve from $+\infty$ to o will give radii vectores of every magnitude from ∞ to 1, and parabolic arcs from ∞ to o. Hence, while the numbers range from ∞ to 1, the parabolic arcs range from ∞ to o. When the number lies between 1 and o, the radius vector representing it is drawn below the axis; its extremity will be found on the loop, and the corresponding arc of the parabola will be negative. Hence the logarithm of a positive number is equal to the logarithm of its reciprocal, with the sign changed; for the magnitude of the parabolic arc depends upon θ, and θ is the same in $\sec \theta + \tan \theta$ as in its reciprocal $\sec \theta - \tan \theta$.

Hence, while the infinite branch of the logocyclic curve from $+\infty$ through R, O, S, to F, may by its radii vectores represent all positive numbers from $+\infty$ to $+o$, the two infinite branches of the parabola will be used in representing the logarithms of positive numbers from $+\infty$ to $+o$; that is, the upper or positive branch of the parabola will be "used up" in representing the logarithms of positive numbers from $+\infty$ to $+1$, and the lower or negative branch of the parabola in representing the logarithms of positive fractional numbers from $+1$ to $+o$. Hence there is no construction by which we can represent negative numbers or their logarithms; therefore such numbers can have no real logarithms.

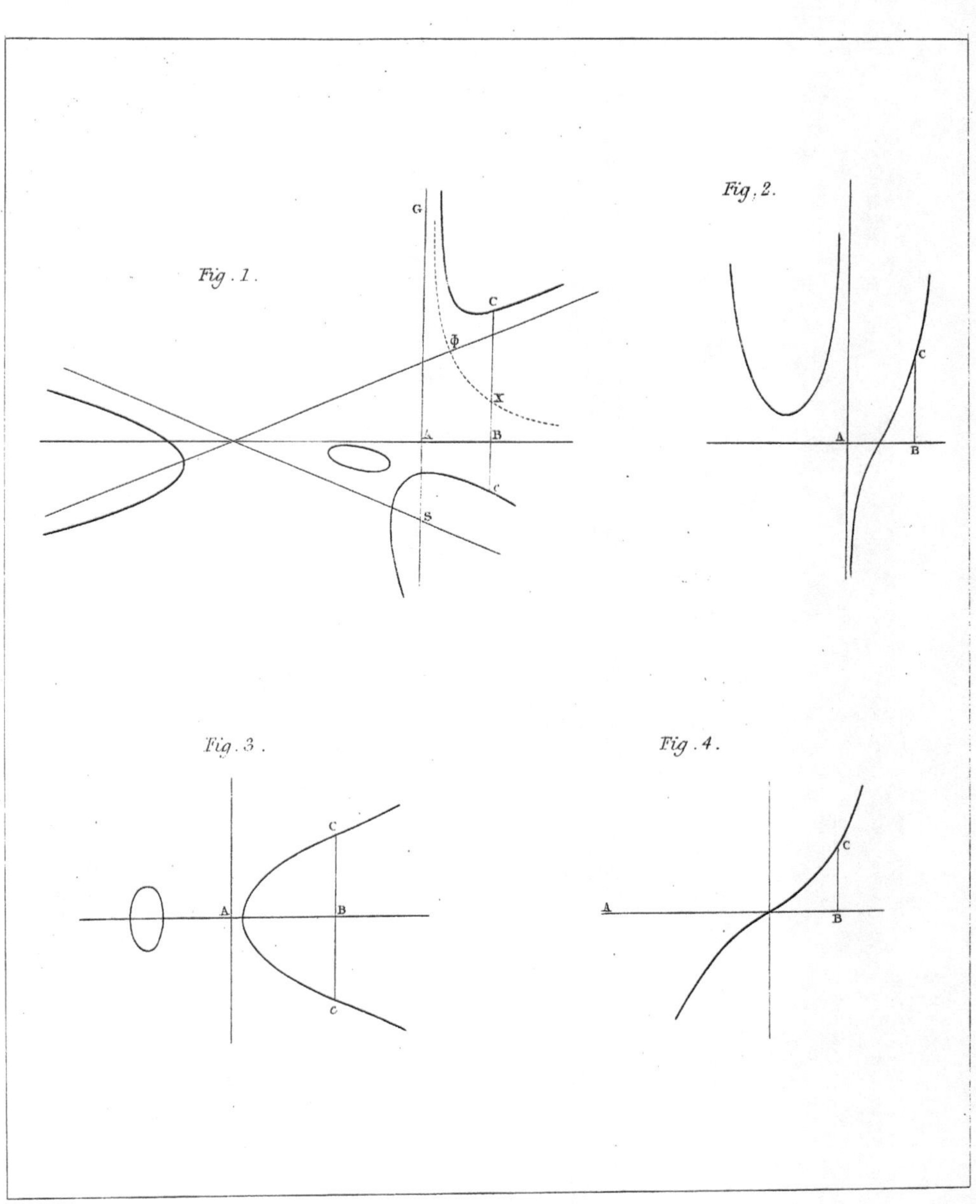

Fig. 1.
G
C
Φ
X
A
B
c
S
Fig. 2.
A
B
C
Fig. 3.
C
A
B
c
Fig. 4.
A
B
C

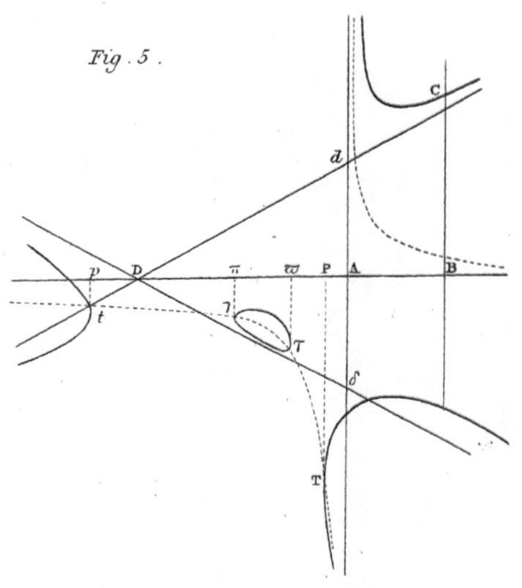

Fig. 5.

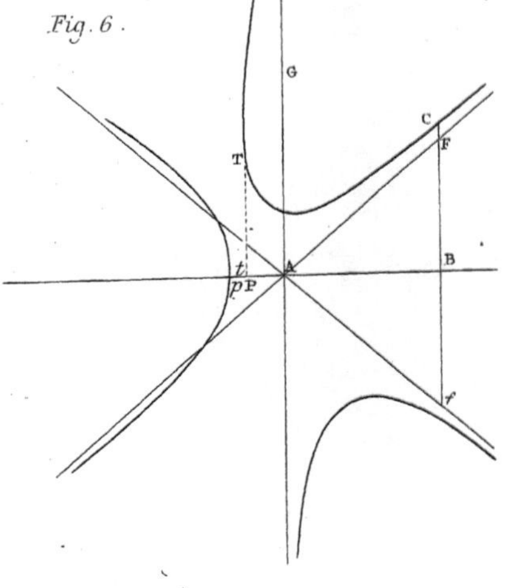

Fig. 6.

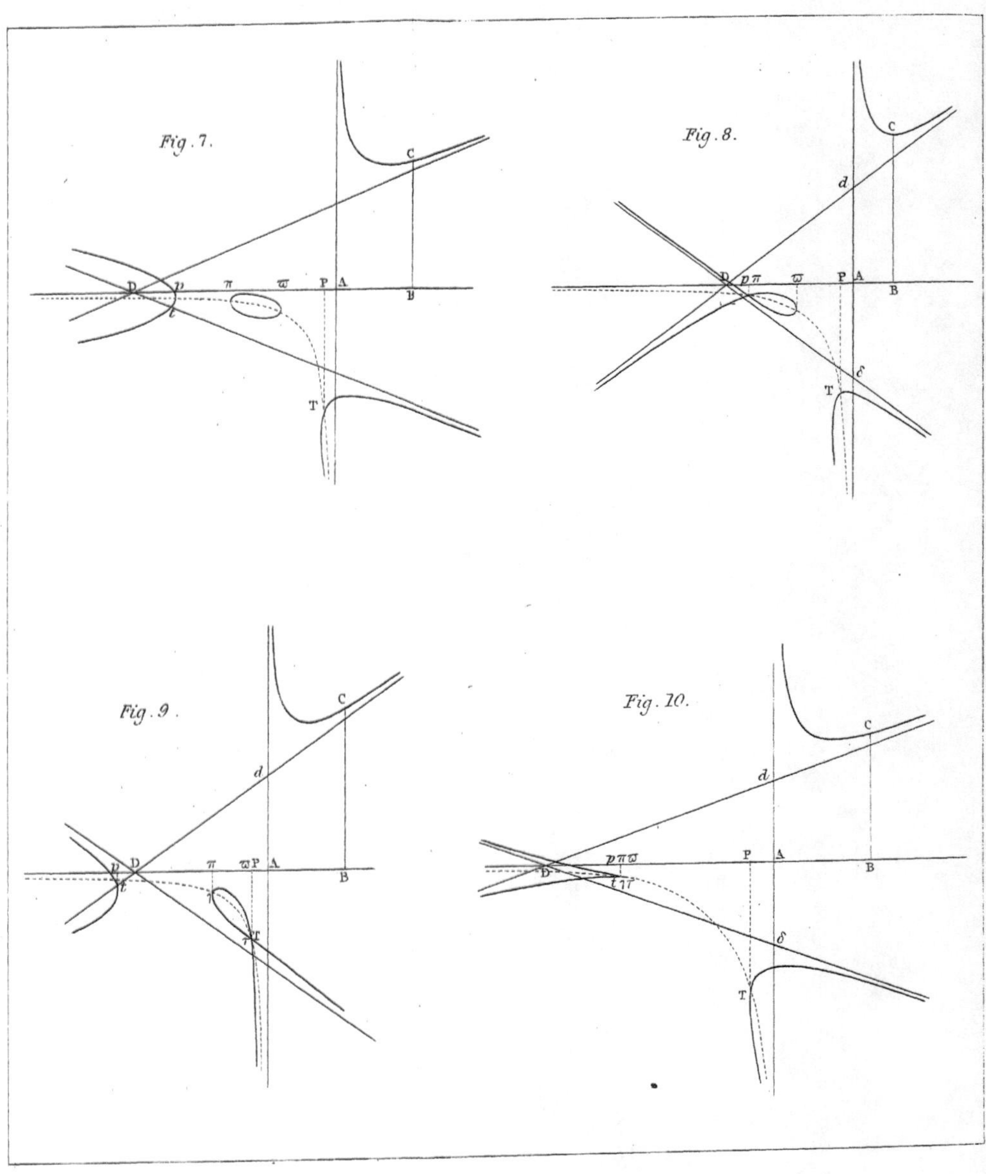

Fig. 7.
Fig. 8.
Fig. 9.
Fig. 10.

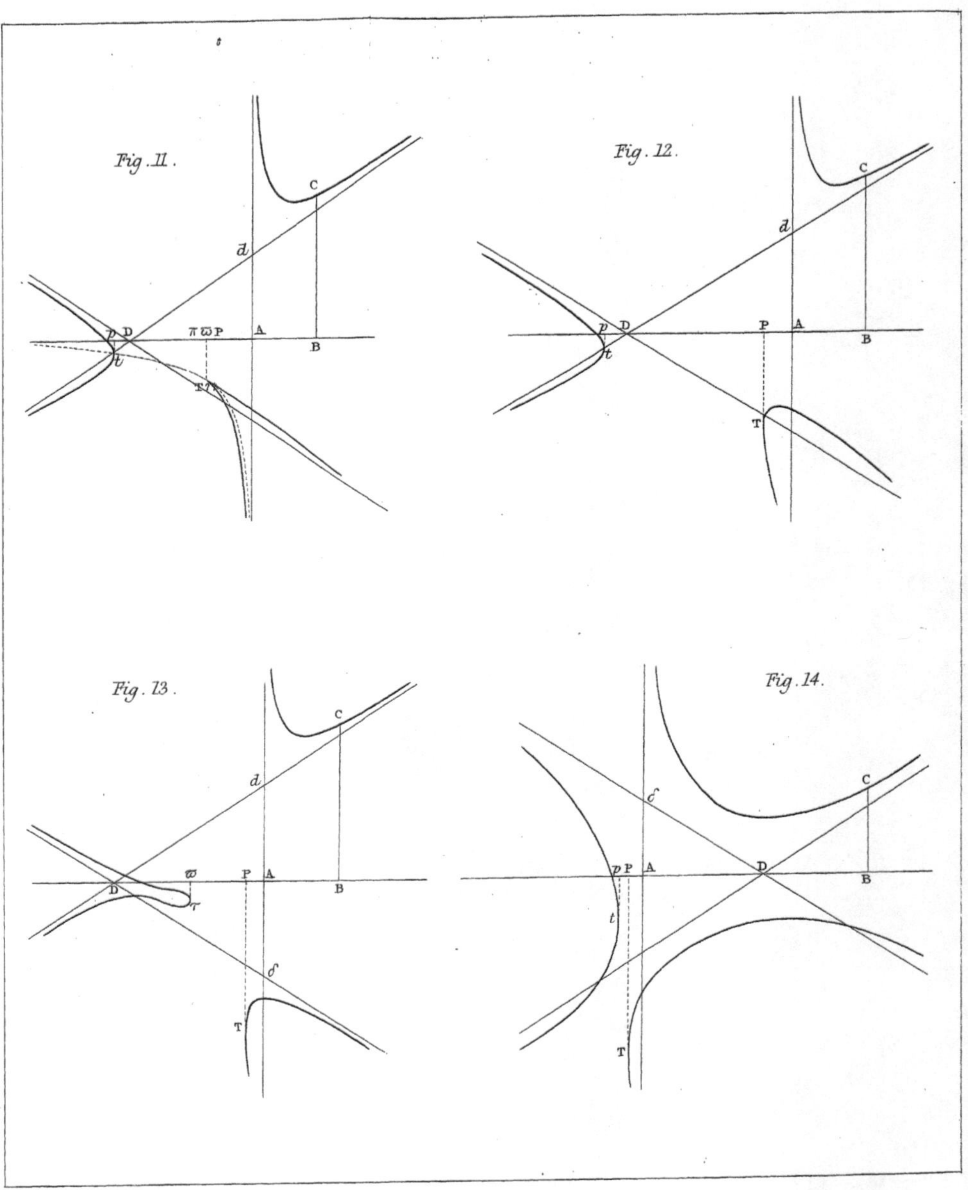

Fig. 11.

Fig. 12.

Fig. 13.

Fig. 14.

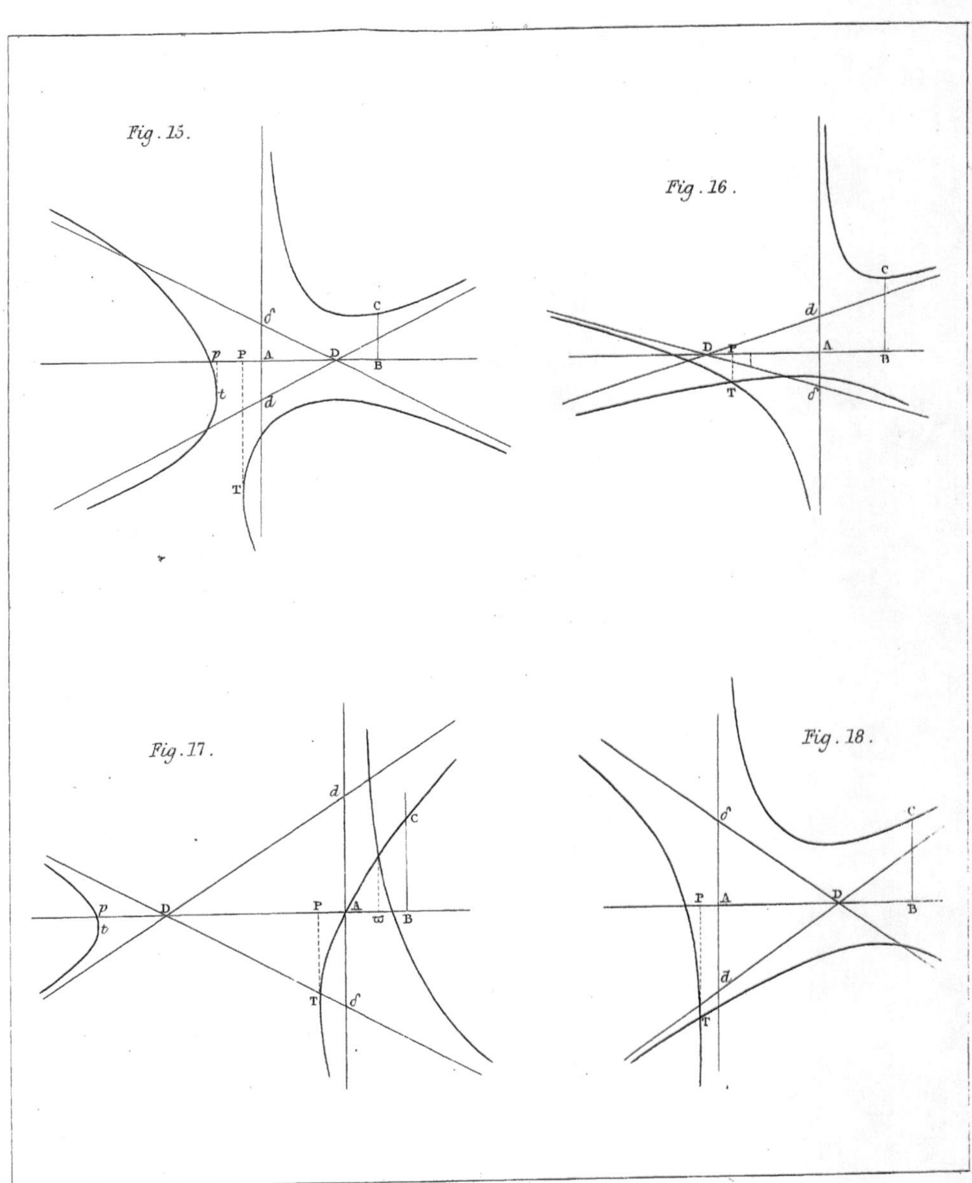

Fig. 15.
Fig. 16.
Fig. 17.
Fig. 18.

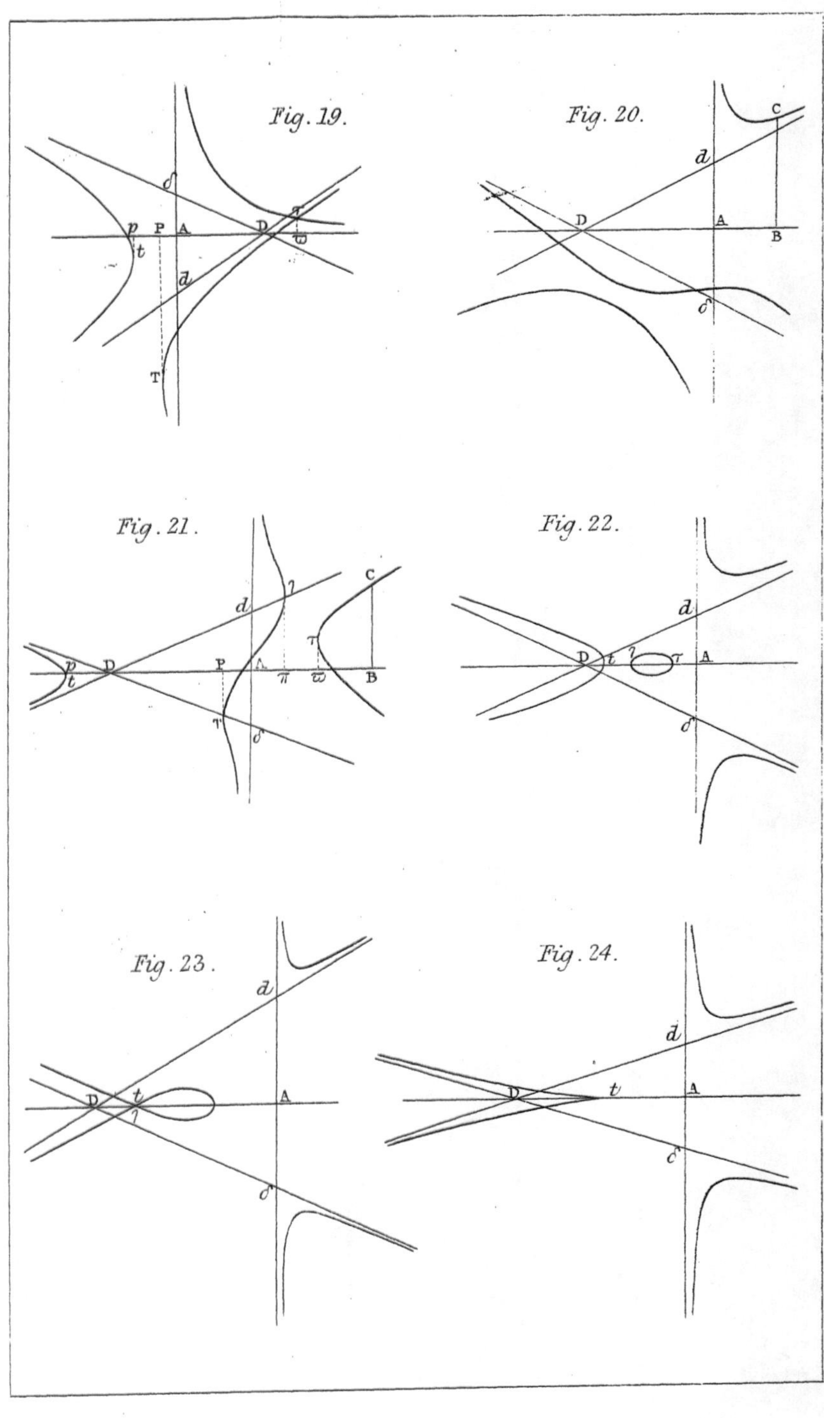

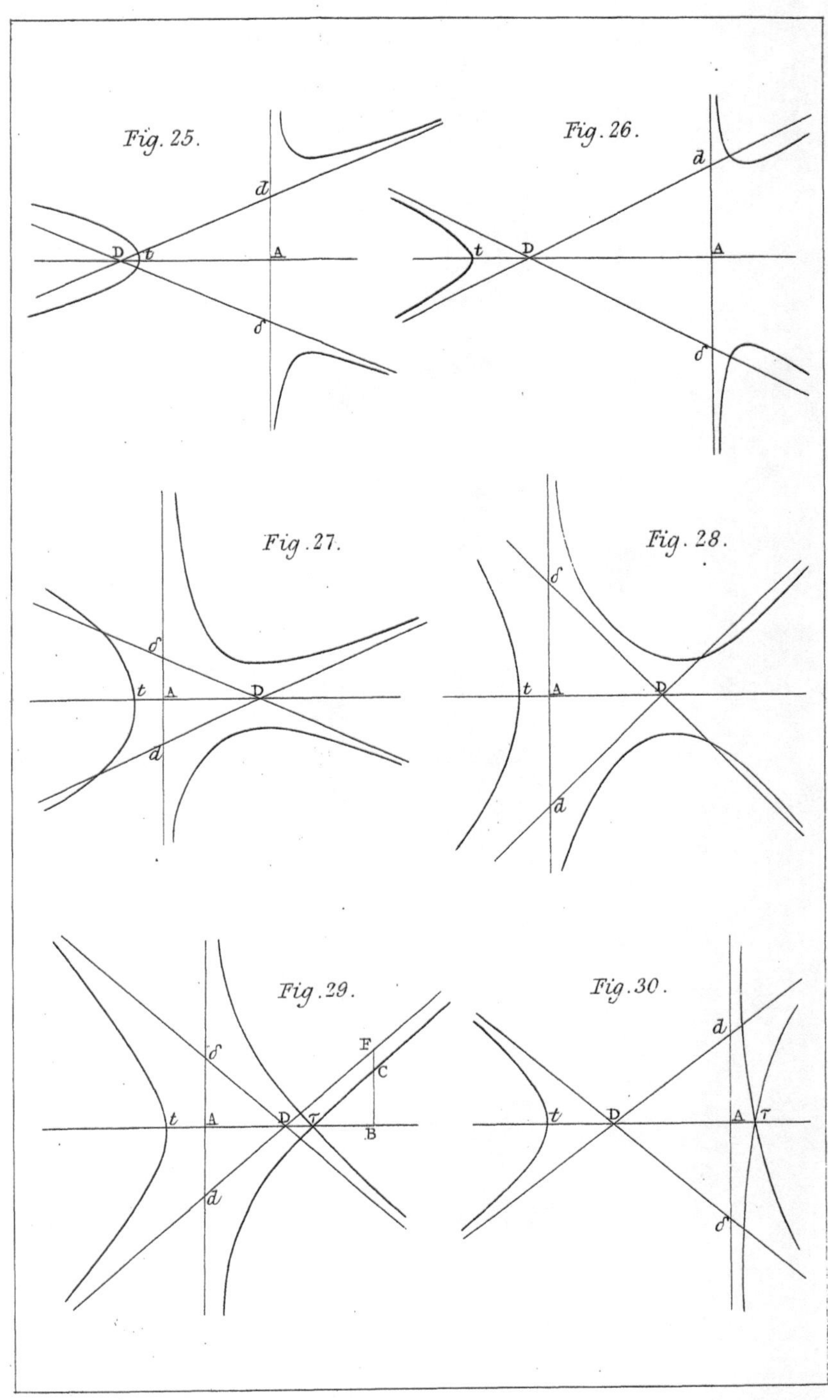

Fig. 25.
Fig. 26.
Fig. 27.
Fig. 28.
Fig. 29.
Fig. 30.

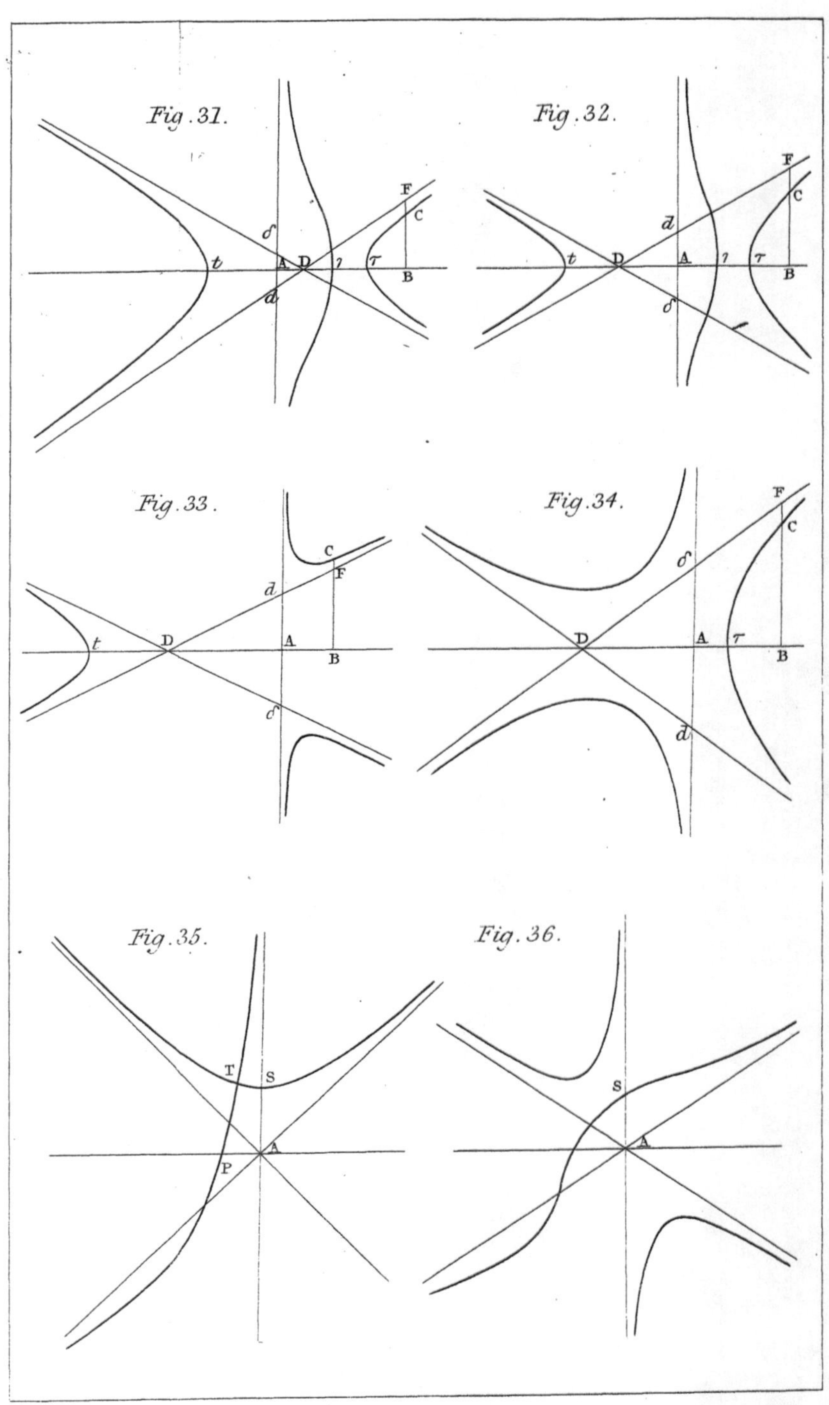

Fig. 31.
Fig. 32.
Fig. 33.
Fig. 34.
Fig. 35.
Fig. 36.

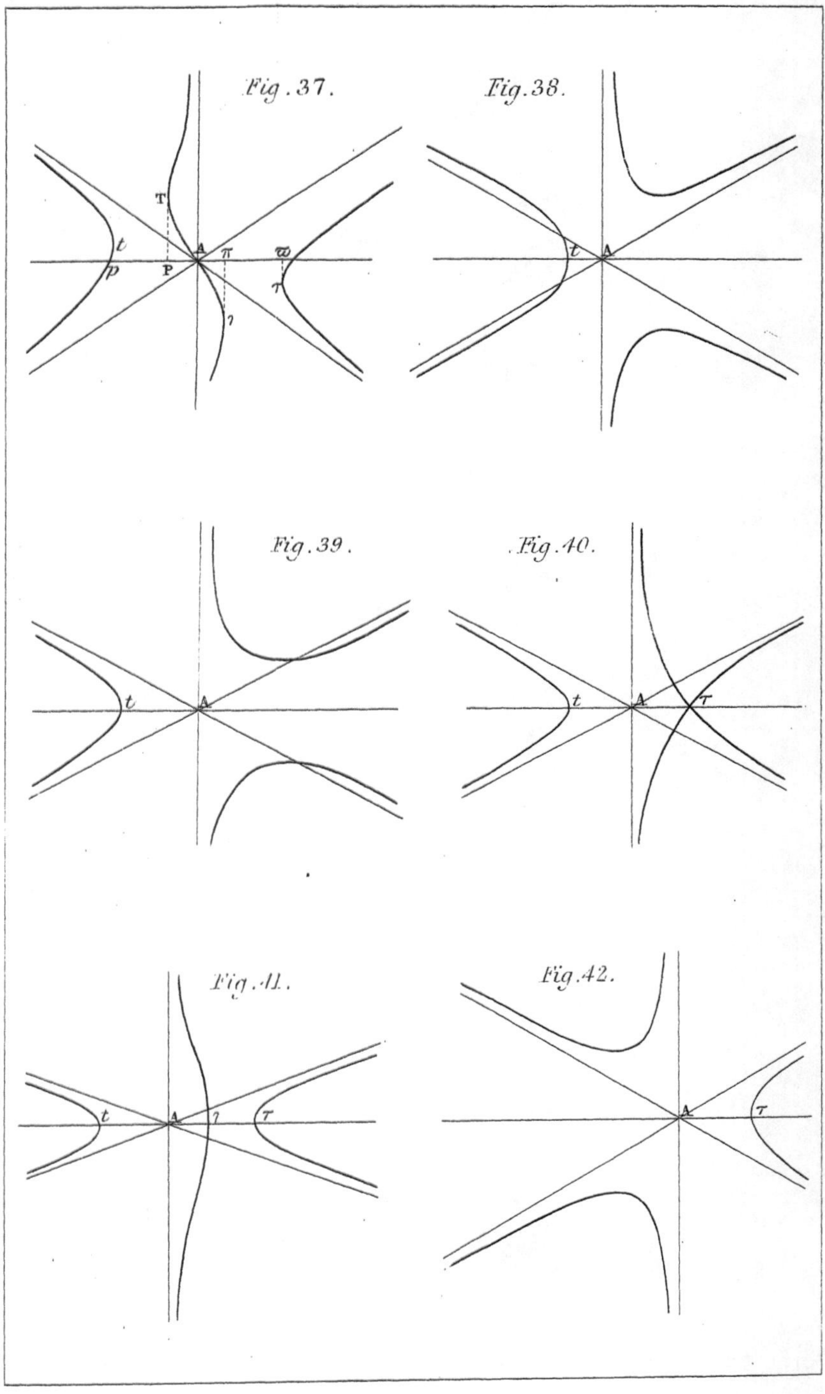

Fig.37.
Fig.38.
Fig.39.
Fig.40.
Fig.41.
Fig.42.

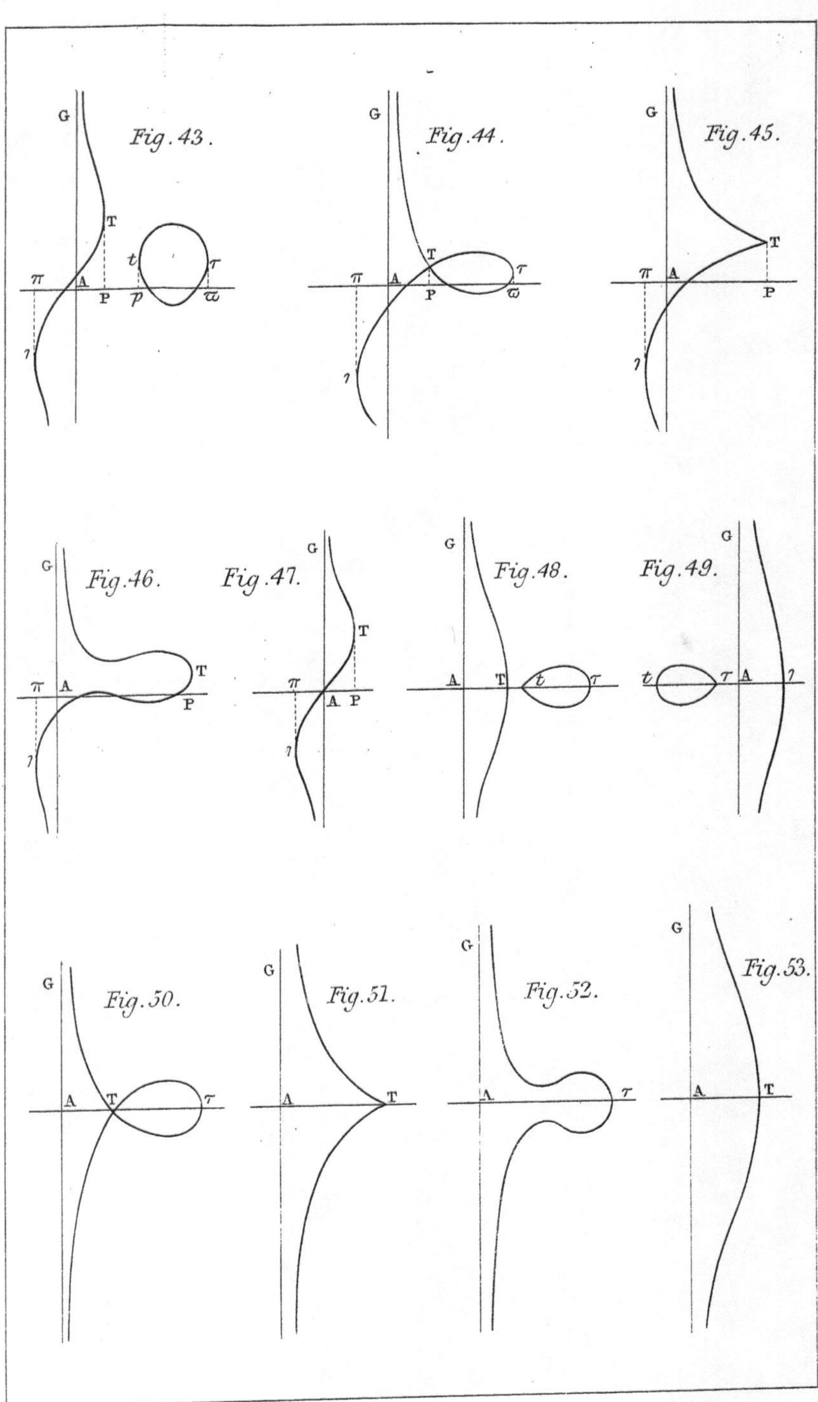

Fig. 43.

Fig. 44.

Fig. 45.

Fig. 46.

Fig. 47.

Fig. 48.

Fig. 49.

Fig. 50.

Fig. 51.

Fig. 52.

Fig. 53.

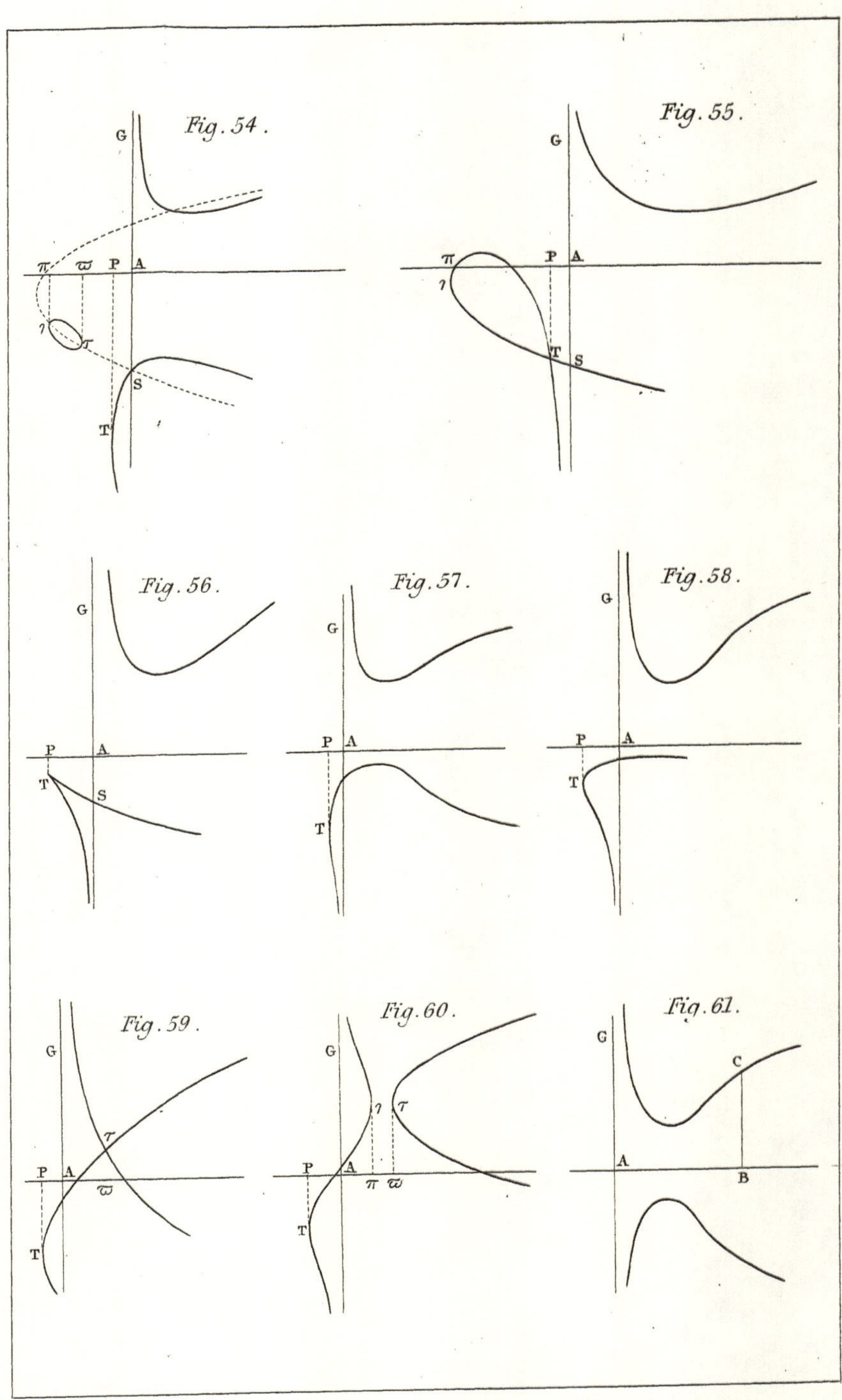

Fig. 54.

Fig. 55.

Fig. 56.

Fig. 57.

Fig. 58.

Fig. 59.

Fig. 60.

Fig. 61.

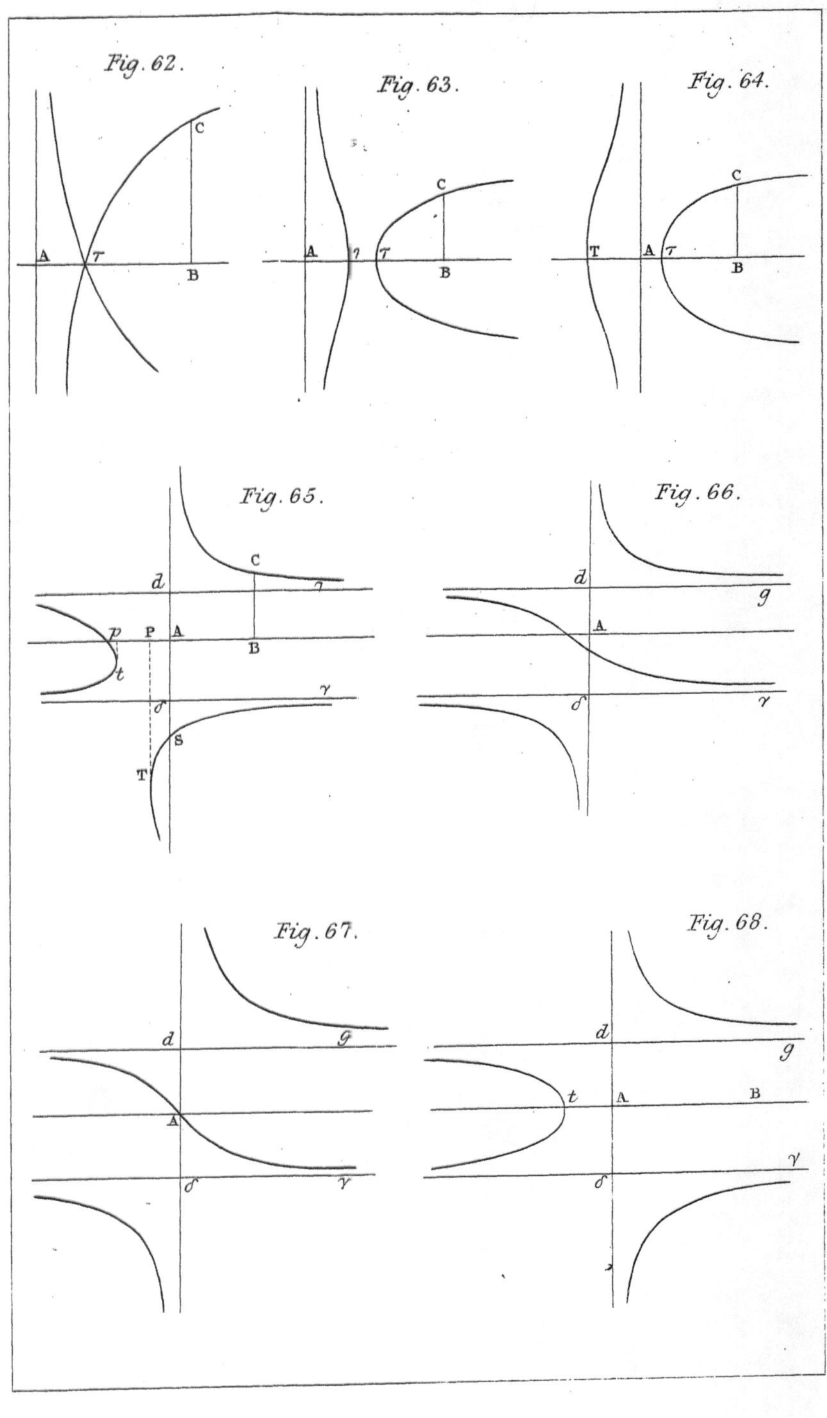

Fig. 62.
Fig. 63.
Fig. 64.
Fig. 65.
Fig. 66.
Fig. 67.
Fig. 68.

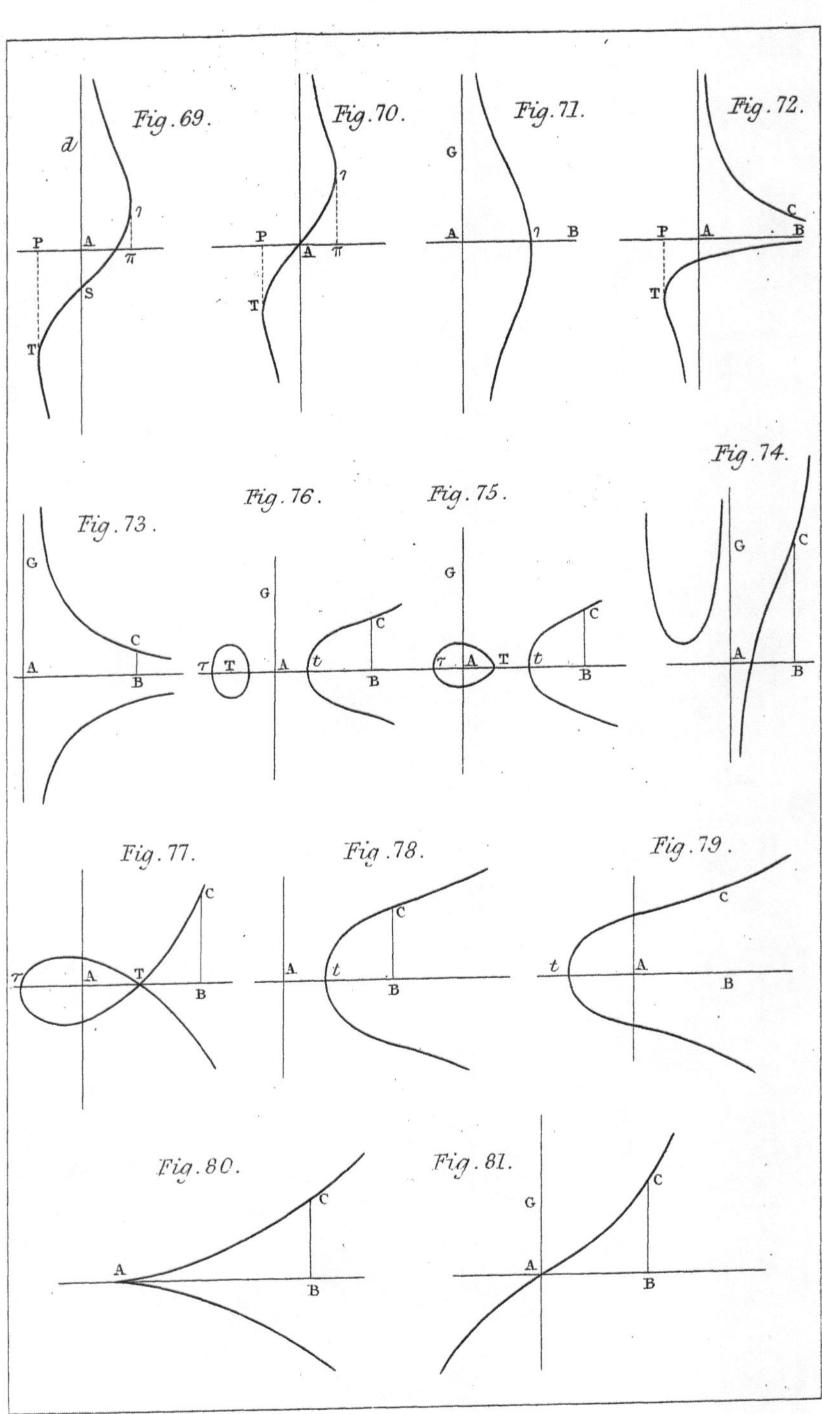

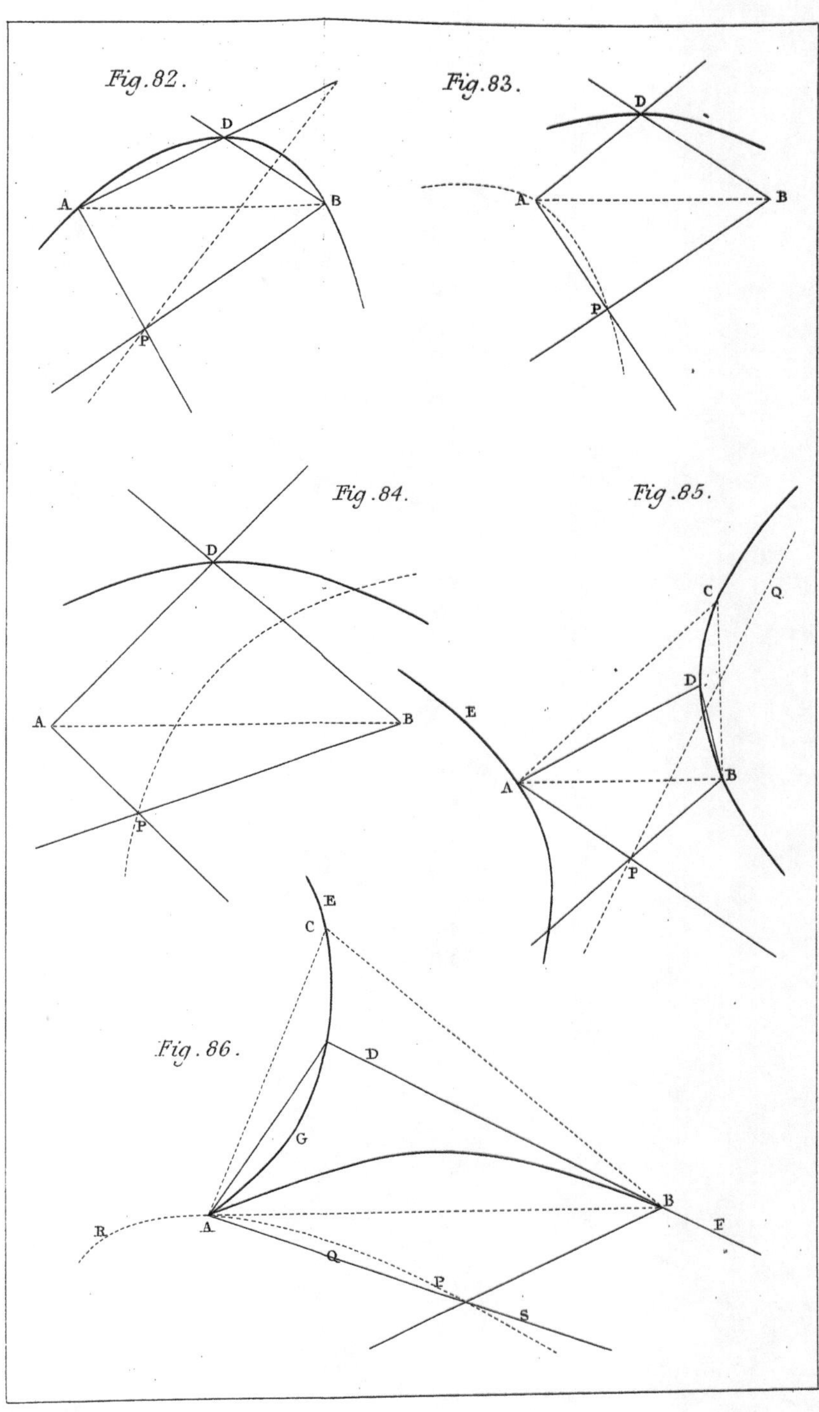

Fig.82.
D
A
B
P
Fig.83.
D
A
B
P
Fig.84.
D
A
B
P
Fig.85.
C
Q
D
E
A
B
P
Fig.86.
E
C
D
G
B
R
A
Q
F
P
S